Identification and Management of Distributed Data

NGN, Content-Centric Networks and the Web

OTHER TELECOMMUNICATIONS BOOKS FROM AUERBACH

Ad Hoc Mobile Wireless Networks: Principles, Protocols, and Applications, Second Edition
Subir Kumar Sarkar, T.G. Basavaraju, and C. Puttamadappa
ISBN 978-1-4665-1446-1

Building Next-Generation Converged Networks: Theory and Practice
Al-Sakib Khan Pathan, Mostafa Monowar, and Zubair Md. Fadlullah (Editors)
ISBN 978-1-4665-0761-6

Game Theory in Communication Networks: Cooperative Resolution of Interactive Networking Scenarios
Josephina Antoniou and Andreas Pitsillides
ISBN 978-1-4398-4808-1

Green Communications and Networking
F. Richard Yu, Xi Zhang, and Victor C.M. Leung (Editors)
ISBN 978-1-4398-9913-7

Handbook on Mobile and Ubiquitous Computing: Status and Perspective
Laurence T. Yang, Evi Syukur, and Seng W. Loke (Editors)
ISBN 978-1-4398-4811-1

Identification and Management of Distributed Data: NGN, Content-Centric Networks and the Web
Giovanni Bartolomeo and Tatiana Kováčiková
ISBN 978-1-4398-7907-8

Intelligent Sensor Networks: The Integration of Sensor Networks, Signal Processing and Machine Learning
Fei Hu (Editor)
ISBN 978-1-4398-9281-7

The Internet of Things in the Cloud: A Middleware Perspective
Honbo Zhou
ISBN 978-1-4398-9299-2

Linear Programming and Algorithms for Communication Networks: A Practical Guide to Network Design, Control, and Management
Eiji Oki
ISBN 978-1-4665-5263-0

Multihomed Communication with SCTP (Stream Control Transmission Protocol)
Victor C.M. Leung, Eduardo Parente Ribeiro, Alan Wagner, and Janardhan Iyengar
ISBN 978-1-4665-6698-9

Next-Generation Mobile Broadcasting
David Gómez-Barquero
ISBN 978-1-4398-9866-6

Optimal Resource Allocation for Distributed Video Communication
Yifeng He and Ling Guan
ISBN 978-1-4398-7514-8

SC-FDMA for Mobile Communications
Fathi E. Abd El-Samie, Faisal S. Al-kamali, Azzam Y. Al-nahary, and Moawad I. Dessouky
ISBN 978-1-4665-1071-5

Security and Privacy in Smart Grids
Yang Xiao (Editor)
ISBN 978-1-4398-7783-8

Security for Wireless Sensor Networks using Identity-Based Cryptography
Harsh Kupwade Patil and Stephen A. Szygenda
ISBN 978-1-4398-6901-7

Transmission Techniques for 4G Systems
Mário Marques da Silva, Americo Correia, Nuno Souto, and Joao Carlos Silva
ISBN 978-1-4665-1233-7

Wireless Sensor Networks: Current Status and Future Trends
Shafiullah Khan, Nabil Ali Alrajeh, and Al-Sakib Khan Pathan (Editors)
ISBN 978-1-4665-0606-0

AUERBACH PUBLICATIONS
www.auerbach-publications.com
To Order Call: 1-800-272-7737 • Fax: 1-800-374-3401
E-mail: orders@crcpress.com

Identification and Management of Distributed Data

NGN, Content-Centric Networks and the Web

Giovanni Bartolomeo

Tatiana Kováčiková

CRC Press is an imprint of the
Taylor & Francis Group, an **informa** business

CRC Press
Taylor & Francis Group
6000 Broken Sound Parkway NW, Suite 300
Boca Raton, FL 33487-2742

First issued in paperback 2019

© 2013 by Taylor & Francis Group, LLC
CRC Press is an imprint of Taylor & Francis Group, an Informa business

No claim to original U.S. Government works

ISBN-13: 978-1-4398-7907-8 (hbk)
ISBN-13: 978-0-367-37996-4 (pbk)

This book contains information obtained from authentic and highly regarded sources. Reasonable efforts have been made to publish reliable data and information, but the author and publisher cannot assume responsibility for the validity of all materials or the consequences of their use. The authors and publishers have attempted to trace the copyright holders of all material reproduced in this publication and apologize to copyright holders if permission to publish in this form has not been obtained. If any copyright material has not been acknowledged please write and let us know so we may rectify in any future reprint.

Except as permitted under U.S. Copyright Law, no part of this book may be reprinted, reproduced, transmitted, or utilized in any form by any electronic, mechanical, or other means, now known or hereafter invented, including photocopying, microfilming, and recording, or in any information storage or retrieval system, without written permission from the publishers.

For permission to photocopy or use material electronically from this work, please access www.copyright.com (http://www.copyright.com/) or contact the Copyright Clearance Center, Inc. (CCC), 222 Rosewood Drive, Danvers, MA 01923, 978-750-8400. CCC is a not-for-profit organization that provides licenses and registration for a variety of users. For organizations that have been granted a photocopy license by the CCC, a separate system of payment has been arranged.

Trademark Notice: Product or corporate names may be trademarks or registered trademarks, and are used only for identification and explanation without intent to infringe.

Visit the Taylor & Francis Web site at
http://www.taylorandfrancis.com

and the CRC Press Web site at
http://www.crcpress.com

Contents

Section V Linked Data

Preface

The motivation for this book was to provide students and future information engineers with a useful collection of Internet standards, technologies, and techniques that derive from research projects for the management of distributed data. We wrote the book during a very exciting period of the Internet. The ubiquitous access of the Internet in daily life by all has allowed the emergence of new sorts of businesses, based on the staggering—and at times puzzling—amount of information on the Internet. Search engines, social networking, online advertising, and online commerce have been built on the aggregation and sharing of personal data. Carriers closely cooperate with content providers to allow a deeper integration between their services and the underlying network, therefore improving the quality of their customers' experience. Governments and public administrations have begun to simplify the way public data are accessed by opening their archives to citizens through the web.

Although several academic courses are offered that discuss data management and networking, few of them focus on the convergence of networking and software technologies for identifying, addressing, and managing distributed data. This book focuses on this convergence—a result of a series of long processes in the history of the Internet in which data management has been thought and rethought many times.

The different layers of the Internet protocol stack already provide different functions as well as useful analogies that provide information engineers with the opportunity and knowledge to design efficient systems.

The best way to read this book is to be aware of some of the recurring themes we deal with in its five sections. The first three themes are explicitly related to data identification, while the rest deal with data management.

What or Where?

The ambiguity between *where* and *what* has haunted the history of the Internet since its early days. Does an identifier give a clear indication of what is being identified? An Internet Protocol (IP) address indicates only the network address of a likely unknown host. The Domain Name System (DNS) is required to deliver a more human-readable and memorable domain name. Other identifiers are intended to provide finer granule identification of resources by name (Uniform Resource Name [URN]) or location (Uniform Resource Locator [URL]), but although they may be meaningful to the human user, they remain unintelligible to the machine.

Flat or Hierarchical Names?

Should names or addresses be issued in a centralized or distributed manner? The widespread DNS is an efficient, hierarchical, and centralized mechanism. Some peer-to-peer networks also provide an equally efficient and scalable lookup system by implementing the Distributed Hash Table (DHT) algorithm, which creates a numeric address space into which decentralized flat names are mapped.

Trustworthiness

Are there inherent mechanisms to ensure that a user accessing a resource can trust its information and the validity of its various parameters, such as its ownership? Today many mechanisms rely on the trustworthiness of the host and not on that of the actual resource being accessed.

Efficient Data Distribution

Multicast backbone (Mbone) was the first attempt at multicasting on the Internet, but several factors—such as the lack of semantics for controlling its channels—made it less appealing and led to solutions based on a multiunicast application level protocol (i.e., HTTP). However, to be effective these solutions require caching mechanisms implemented in specialized network appliances in the Internet. Novel content-centric paradigms aim to simplify and improve these solutions, introducing caching directly into the network layer.

Representational State Transfer

The Representational State Transfer (REST) architectural style was formally defined more than 10 years ago, but only recently has its unifying power been fully understood. REST applies to every kind of resource and defines uniform addressing mechanisms, common semantics for operations, flexible ways of transferring resource representations, and support for caching. Data management applications built on REST are efficient, avoid duplications, and improve performances.

Resource-Oriented Models versus Representation-Oriented Markup Languages

Extensible Markup Language (XML) and more recently JavaScript Object Notation (JSON) are open and extendible standard formats adopted by several technologies to model, store, and transmit data through the wire. These formats are representation oriented, however, as they rely on concrete encoding. Resource Description Framework (RDF) models, on the contrary, in the form of graphs, are independent in how they are encoded. RDF is a versatile resource-oriented data model rather than a specific serialization format.

Linking Data

Data linkage has been reinvented several times in data management applications. RDF has ultimately provided a native mechanism allowing clients to link data the way pages are linked on the traditional web. An ideal "giant RDF graph" is under construction, and there are expectations of how it could be used to answer queries previously thought impossible.

Due to the extensive nature of the subject, this book, far from being exhaustive, provides only hints of a limited number of relevant technologies. But we think this is a good starting point. Indeed, in writing the book we have put more emphasis on illustrating how the existing Internet stack already provides all the necessary functions to handle and distribute data rather than describing new specialized layers. Some specific technologies may be obsolete. However, we believe that the themes we address are relevant in that they have been there since the origins of networking and software engineering, and it is likely they will be there for a long time into the future.

From networks addressing basics to controversial questions on equivalence and identity, our long journey can now begin.

Acknowledgments

This book would not have been possible without the valuable input, contributions, feedback, and even criticism of our colleagues and friends. We thank all the people who have been involved in the development of this manuscript.

Giovanni acknowledges the extraordinary people he has had the honor of working with during his eight years with the networking group in the Department of Electronic Engineering at the University of Rome Tor Vergata. In particular, he wishes to thank Stefano Salsano and Matteo Cancellieri, who provided outstanding guidance in the understanding of content-centric networking as well as other precious related material.

Tatiana acknowledges Pavel Segec, from the Department of InfoCom Networks at the University of Žilina, Slovakia, for their long-term and fruitful collaboration in the area of next-generation networks.

We owe a special thanks to our friends Dion Drislane and Uma Arunachalam, who kindly offered to clean up our nonnative English.

We also wish to acknowledge all the editorial staff at Taylor & Francis, particularly Richard O'Hanley, who first contacted us with the proposal to write this book and then maintained remarkable patience in light of our many delays in delivering the final manuscript.

About the Authors

Giovanni Bartolomeo earned his Laurea degree in software engineering in 2004 from the Università degli Studi di Palermo; he won a Nortel Networks–funded prize that supported his research activity during his thesis development. As a research collaborator with the Consorzio Nazionale Italiano per le Telecomunicazioni, he participated in several European Union–funded research and development projects and contributed to the World Wireless Initiative "Book of Vision 2008." Between 2008 and 2010, he served as an expert on the European Telecommunications Standards Institute (ETSI) Human Factors Technical Committee. Currently, Bartolomeo is a technical officer at the Italian Ministry of Justice and is involved in different standardization efforts at the Organization for the Advancement of Structured Information Standards (OASIS).

Tatiana Kováčiková earned her MSc degree in telecommunication engineering from the University of Transport and Telecommunications in Žilina, former Czechoslovakia. She earned her PhD in telecommunication systems from the same university in 1995. In 2004, she was appointed associate professor of information and management systems at the University of Žilina. From 1984 to 1988 she worked at Slovak Telecom, and since then she has worked at the University of Žilina, Slovakia. In June 2010 she was appointed head of the Department of InfoCom Networks there. Kováčiková's research interests include Internet Protocol and next-generation network architecture, protocols, and applications, on which she has been involved in several national and international research projects. As a leader of a research group, she received the 2003 Siemens Award in the field of Internet Protocol telephony. Since 2002, Kováčiková has been actively involved in the European Telecommunications Standards Institute (ETSI) GRID, Telecoms and Internet converged Service and Protocol for Advanced Networks (TISPAN), Human Factors, and User Group Technical Committees.

SECTION I
Naming and Addressing on the Internet

A number of standards, specifications, and techniques that have major significance in the field of data-centric networks are described throughout the book. What emerges is that, far from being diverse and uniquely different, there are some very significant similarities and relationships between these different specifications. These similarities are not coincidental but derive from a solid foundation of fundamental Internet standards that have proven to be very flexible in the way that they can be applied and adapted.

This section introduces the most fundamental of the Internet naming and addressing standards on which so many later specifications were able to build. The fundamental simplicity of many of these standards has allowed them to be built upon in both hierarchical and evolutionary ways. So, for example, the most widely exploited Hypertext Transfer Protocol (HTTP) is reliant on the Transmission Control Protocol (TCP)/Internet Protocol (IP) for routing messages and establishing connections and on Domain Name System (DNS) for domain name resolution. Many other specifications apply the structures and techniques inherent in HTTP to other contexts than the one for which HTTP was designed. These relationships and similarities offer very real and practical benefits in terms of the way existing skills, algorithms, and software can be used and adapted to allow newer emerging ideas to be rapidly exploited and evolved.

1
IP Addresses

Even if today it is common to think of the Internet in terms of its human-readable domain names and host names, the latter still leverage the Internet Protocol (IP), the addressing mechanism introduced in the late 1970s by Steve Cocker and Jon Postel. An IP provides the connectionless datagram service that today is at the heart of modern internetworking. It governs how hosts, networks, and subnetworks are identified across the Internet and how packets are routed through them. Remaining agnostic to the intricacies of the underlying communication layers, it represents the first abstraction available to network engineers and software developers. It is thus the ideal starting point to begin the study of any identification mechanism available on the Internet.

Internet Protocol, Version 4

The Internet Protocol, version 4 (IPv4), addresses are 32-bit binary numbers assigned to hosts and networks in the Internet or in Local Areas Networks (LANs) with Transmission Control Protocol (TCP)/IP technology. IPv4 addresses are represented in a human interface in the more memorable "dotted decimal notation," where the sequence of 32 bits is broken down into four octets (i.e., sets of 8 bits), which are converted to their decimal value, with the leftmost bit being the least significant bit (LSB) and the rightmost one the most significant bit (MSB). Thus, an IPv4 address (e.g., 192.168.10.254 or 10.0.244.1) appears as a sequence of four 8-bit numbers.

The IPv4 address space contains over 4 billion different addresses. However, the protocol imposes certain restrictions on how these addresses are allocated and how they are managed and reserved for special purposes. All of these restrictions place a practical limit on the total number of available addresses.

An IPv4 address (henceforth IP address) actually consists of two identifiers: (1) the network identifier and (2) the host identifier. The bits of an IP address are divided into two sequences: the leftmost bits of the IP address are assigned to the network identifier, and the rightmost ones are assigned to the host identifier. Some combinations of these bits are reserved for special purposes; in particular, the host identifier bits cannot all be set to 1 or all set to 0.

IP Classes

The original IP specifications (RFC 971), dated September 1981, introduced the concept of *address classes* to clearly differentiate among five different sizes of networks that are connected to the Internet.

The first four leftmost bits of an IP address define the class to which the IP address belongs (Figure 1.1). Class A IP addresses have their first leftmost bit equal to zero; the network identifier is 8 bits long and is determined by the 8 bits of the first octet. As the first bit is always set to zero, there are only 128 different possible identifiers available. Since 127 is reserved to refer to the loopback mode and 0 is not used, valid class Network identifiers can range only from 1 to 126. The remaining part of the IP address, that is, the remaining three octets, is entirely assigned to the host identifier, providing 16,777,214 different host addresses (having excluded the two combinations of bits corresponding to all zeroes, reserved for the address assigned to the network itself, and all ones, reserved for the broadcast address for all the hosts contained in the network). Class A networks are therefore

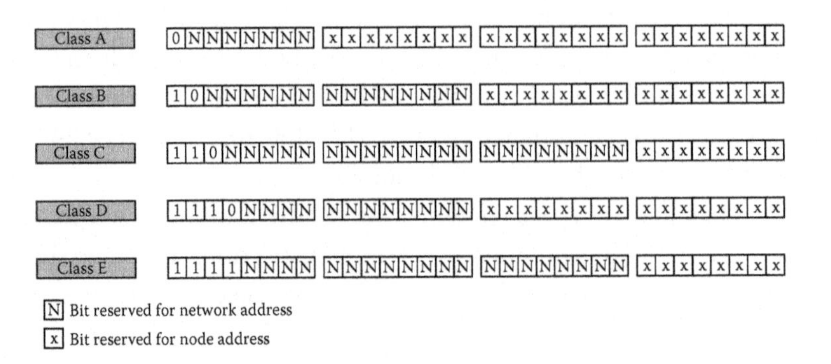

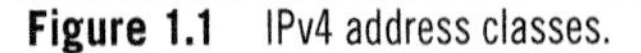

Figure 1.1 IPv4 address classes.

very big networks in terms of the number of hosts they can contain. Many of these addresses belong to networks owned by international organizations (e.g., some well-known commercial companies, government or educational organizations). A complete list of public Class A address assignments (e.g., Class A network identifiers assigned in the context of the Internet) is available at the Internet Assigned Numbers Authority (IANA) registries, accessible online.*

Class B network addresses are characterized by having the first leftmost bit set to 1 and the second leftmost bit set to 0 (the decimal value of the first octet therefore ranges from 128 to 191). Class B network identifiers are 16 bits long, but with the first two leftmost bits already set to 10 the number of possible different network identifiers is reduced to 16,384. The remaining two octets are for the host address and give 65,534 different host addresses (as usual, the combinations of all 1s and all 0s, reserved for the broadcast address and the network address, are excluded).

The third class, Class C, has the first three leftmost bits set to 110 (the decimal value of the first octet therefore ranges from 192 to 223). The length of the network identifier is 24 bits long, but with the first three leftmost bits already set to 110 the number of possible different network identifiers is reduced to 2,097,152. The number of different addressable network is therefore over two million. The benefit that this wide network addressing capability brings is, however, compensated for by a poor host addressing capability. As only 8 bits (one octet) are available to identify the network's host addresses, only 254 network addresses (as 0 and 255 are reserved addresses) can be assigned. In a Class C network, therefore, the first three leftmost octets define the network identifier, whereas only the rightmost octet defines the identifier of the host.

Class D addresses have the first four leftmost bits set to 1110 and are reserved for multicast groups (RFC 1112, Deering 1989). Contrary to unicast, that is, the traditional point-to-point communication paradigm originally thought for the Internet protocol, multicast routers allow (best effort) one-to-many communication among multicast group members, a capability typically required in media streaming applications. After the 4 bits used to identify class D addresses, the

* The IANA IPv4 Address Space Registry, http://www.iana.org/assignments/ipv4-address-space/ipv4-address-space.xml.

remaining 28 bits represent the group identifier; therefore, there can be up to 228 (about 260 million) multicast groups. IP hosts report their host group memberships to local multicast routers using the Internet Group Management Protocol (IGMP). To receive and process packets only from subscribed groups, the network interface of each node is equipped with a multicast filter typically implemented in the hardware. The multicast routers maintain multicast paths, which ensure that few packets (ideally only one per link) are routed through the network and optimally delivered to subscribers.

The original multicast model, defined in RFC 1112 (Deering 1989), was conceived as an open group model, where any host on the Internet could subscribe to a multicast group. This openness, however, creates the practical problem of access control: there is currently no way to prevent others from using the same multicast group for purposes other than the one defined by the first subscribers. This makes multicast less appealing for data and content delivery because this lack of control is of much greater concern from a commercial perspective than the efficiency benefits that multicast brings compared to multi-unicast solutions.

The most common persistent multicast groups are currently in the range 224.0.0.0–224.0.0.255 (e.g., 224.0.0.1, all the multicast enabled hosts in the local network) and are mainly used for administration and local network maintenance. Other groups may be dynamically created (and destroyed), and in general, their members span across different networks on the Internet. Due to the need for special hardware, or special functionalities, multicast enabled nodes usually form "islands." The multicast backbone (Mbone) project (Savetz, Randall, and Lepage 1996) in the early 1990s attempted to connect these islands by tunnels that transmit multicast packets encapsulated into unicast packets. More recent research projects have now made Mbone obsolete.

Finally, Class E addresses have the first four leftmost bits set to 1111 and are reserved for future use.

Subnetting

There is a large discrepancy between the possible number of host identifiers in Class B networks and Class C networks. If the number of hosts in a single organization to be directly connected to the Internet just exceeded 254, say 260, then, according to the Class mechanism defined in RFC 791 (Postel 1981), it would need a Class B network, which actually offers the capability to address more than 65,000 hosts. The organization would therefore waste tens of thousands of public IP addresses. Furthermore, if all host identifiers in a Class B (or even in a Class A) network were to be organized in a flat addressing space (assuming this would be physically possible), then there would be a single broadcast domain and the IP packet routing mechanism would be extremely slow for the hosts in that network. Therefore, there is a need to allow a certain degree of organization of IP addresses within each single class of addresses, splitting the flat address space provided by the class mechanism into different broadcast domains.

The method currently used for splitting large classes of addresses is called subnetting. The idea behind subnetting is quite simple and consists of extending the network identifier by using some bits taken from the host identifier. This is possible using local routers configured to indicate which bits of the host identifiers have to be taken as the subnet identifier. This indication is provided to the local router using a subnet mask. A *subnet mask* is a sequence of 32 bits in which the first n-leftmost bits are set to 1 and the remaining bits are set to 0.

For example, in a Class B network, it would make sense to use the first 3 bits of the host identifier to split the host address space into 8 (as 8 is the number of different subnet addresses made possible using only 3 bits) different subnetworks, each of them with 8,190 ($2^{13} - 2$) different hosts. To this end, a corresponding subnet mask would contain a sequence of 19 leftmost bits set to 1 followed by 13 bits set to 0. Subnet masks are usually expressed with the same dotted decimal notation used for IP addresses. The subnet mask always begins with 255. In the previous example, our subnet mask would be written in decimal dotted notation as 255.255.224.0.

The local router provided with this mask simply performs a bitwise operation, a logical AND between the IP address contained in the destination field of an IP packet (e.g., 140.32.66.5) and the subnet mask. The first two octets, corresponding to the network identifier of the target IP address, are unchanged and correctly identify the network address. The first three bits of the third octet are then considered, and the following bits are discarded (as the logical AND with 0 as operand just produces [i.e., gives] the logical value 0). These first three bits represent the local identifier of the subnet to which the target host belongs, in our case 010 in binary notation, which is 2 in decimal. The router can then route the packet on the port corresponding to subnet number 2. The host address is given by the remaining rightmost bits (including those of the fourth octet), having set the leftmost three bits of the third octet to zero.

A subnet mask could be expressed either using the dotted decimal notation or in a notation called a *classless interdomain routing* (CIDR) *notation* (RFC 4632, Fuller and Li 2006), aka *network prefix*. The network prefix consists of a slash symbol after the IP address, followed by the number of consecutive 1s in the subnet mask. In our example, it would be 140.32.66.5/19 (Figure 1.2).

Three ranges of IPv4 addresses (Table 1.1) are explicitly reserved for nodes of local networks that do not need to be directly connected to the Internet (RFC 1918, Rekhter et al. 1996). Through a Node Auto Term (NAT) that is assigned. These addresses are able to initiate outgoing connections to other nodes (typically Web servers, Post Office Protocol [POP] servers, and File Transfer Protocol [FTP] repositories), passing through a Network Address Translation (NAT) node that has a public IP address. Outside the local network, however, these addresses are not perceived as IP nodes. In fact, the NAT receives any outgoing packet from nodes in the local network, changes the source address (and port if necessary) to its own, adjusts the packet's checksum, and routes the packet to the destination. The NAT also creates an association between the original source and the destination so that it will be able to route incoming packets from that destination to the original source.

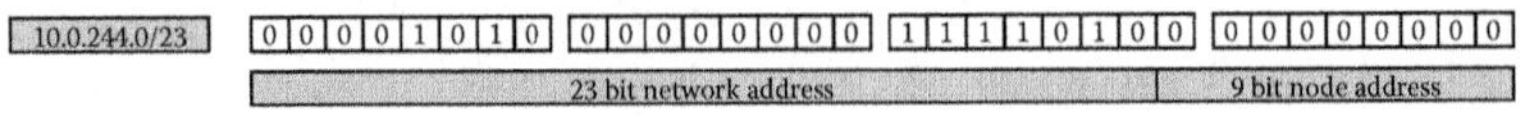

10.0.244.0/23 | 0 0 0 0 1 0 1 0 | 0 0 0 0 0 0 0 0 | 1 1 1 1 0 1 0 0 | 0 0 0 0 0 0 0 0

23 bit network address | 9 bit node address

Figure 1.2 CIDR notation for network 140.32.66.5/19.

Table 1.1 Reserved IPv4 Addresses for Private Networks

RESERVED IP ADDRESSES	RANGE	NUMBER OF HOSTS
10.0.0.0/8	10.0.0.0–10.255.255.255	16,777,216
172.16.0.0/12	172.16.0.0–172.31.255.255	1,048,576
192.168.0.0/16	192.168.0.0–192.168.255.255	65,536

Table 1.2 Special IPv4 Addresses

LEFTMOST BITS (NETWORK ID)	RIGHTMOST BITS (HOST ID)	DESCRIPTION
All bits set to 1 (one)	All bits set to 1 (one)	Broadcast address of the local network
Network ID	All bits set to 1 (one)	Broadcast address of network ID
Network ID	All bits set to 0 (zero)	A network identifier
All bits set to 0 (zero)	All bits set to 0 (zero)	Localhost (used in DHCP)
All bits set to 0 (zero)	Host ID	A host identifier in the local network
$(01111111)_2 = (127)_{10}$	Any value	Loopback addresses

Special IP Addresses

Table 1.2 illustrates some special uses for various combinations of bits. If the rightmost bits of the host address are all set to 1, the address is considered to be a broadcast address. An IP packet sent to a broadcast address is delivered to any host of the network identified by the left-most bits.

If all the leftmost bits are also set to 1, then it is assumed that the target network is the local one (i.e., the one where the packet is generated). If all the leftmost bits (but not the rightmost bits) are set to 0, then the IP address is understood as a host identifier in the local network. If all the rightmost bits (but not the leftmost bits) are set to 0, the host address is not used and the whole IP address is considered to be the IP address of the network. (In practice, it coincides with the network identifier.) The combination corresponding to all 32 bits being set to 0 is reserved for the localhost and is used as a source address only in bootstrapping (with the host requesting an address, as it happens in the Dynamic Host Configuration Protocol).

IP addresses with the first octet made of one 0 followed by seven 1s (i.e., 127.x.y.z in dotted decimal notation) are reserved for loopback and can be used as source as well as destination addresses. Packets sent to loopback addresses are not really routed through the network; rather, they are considered self-incoming packets. Even if they are sometimes colloquially referred to as “localhost,” loopback addresses should not be confused with it, as they serve a very different function (RFC 5735, Cotton and Vegoda 2010).

Internet Protocol, Version 6

With the rapid growth of Internet usage, the practical restrictions on the actual number of different IPv4 addresses that can be allocated has led to recurring alerts about the lack of IP addresses that will be available for future networks. The emergence over time of alternative solutions based on IPv4, NATs, and virtual servers has helped to alleviate the critical lack of IPv4 addresses, with the result that IPv4 is still the most commonly used network protocol. Use of these techniques, however, is controversial as they violate the principle of network transparency, according to which packets should flow totally unaltered (also in their header) from source to destination (RFC 2775, Carpenter 2000). Thus, according to many authors the problem has been delayed, not solved, and its ultimate solution consists in the replacement of IPv4 protocol with its successor, IPv6.

The IPv6 protocol, which was primarily thought to be the ultimate solution to the increasing need for Internet addresses, uses 128-bit addresses theoretically being able to reference up to 2,128 (approximately 3.4×10^{38}) different addressable entities. Actually, this solution is a combination of two different proposals presented by Deering and Francis (1993) and was originally known as Simple Internet Protocol Plus (SIPP; RFC 1710 [Hinden 1994], not to be confused with the Session Initiation Protocol [SIP]). SIPP, however, used only 64-bit addresses, which were considered too short considering the increasing need for addresses. Therefore, IPv6 was standardized with 128-bit addresses. Over the years there have been many creative attempts to highlight how the number of available 128-bit numbers is so large that it would be practically impossible to run out of them. Two colorful examples that have been used are that the average number of IPv6 addresses per squared meter is 7×10^{23} across the whole surface of the earth (including oceans and other water surfaces) and that only 1 percent of the total number of IPv6 addresses would have been allocated if one IPv6 address had been assigned every millisecond since the creation of the universe (15 billion years ago) (Tanenbaum 2007). Table 1.3 gives a more precise comparison of the different orders of magnitude of the two address spaces.

As with IPv4, however, IPv6 addresses are not uniformly allocated, and large portions of the total potential address space are not

Table 1.3 Number of Different IPv4 and IPv6 Addresses

PROTOCOL	BITS	NUMBERS OF DIFFERENT ADDRESSES
IPv4	32	4,294,967,296
IPv6	128	340,282,366,920,938,463,463,374,607,431,768,211,456

used. More details of the allocation of IPv6 addresses are given after the following explanation of IPv6 and DNS integration.

IPv6 has not currently been widely accepted. Probably an important reason that software developers and computer users have been reluctant to adopt IPv6 addresses is because of the address notation, which is definitely less memorable than the dotted decimal notation used in IPv4. Actually, IPv6 addresses are written as eight groups of four hexadecimal digits ("nibbles"), each group representing two bytes:

8000:0000:0000:0000:006B:C28D:1F2F:1276

This notation is better suited to be understood by machines than by human users. However, some shortcuts can be used to reduce the length of this address: for example, adjacent groups of four consecutive zeros can be omitted by replacing them with a couple of colons:

8000::006B:C28D:002F:0276

Also, zeros at the beginning of a group of nibbles can be omitted:

8000::6B:C28D:2F:276

But even in this form, the address is definitely more difficult to remember than a traditional IPv4 address expressed in dotted decimal notation.

As in IPv4, the IPv6 address space is divided into partitions, and each partition is identified by the first leftmost bits in the binary representation of the IPv6 address. Examples of these partitions include provider-based addresses (whose binary 128-bit string representation begins with 010), geographical-based addresses (beginning with 100), and multicast addresses (beginning with the prefix 1111:1111). Old IPv4 addresses also represent a partition of IPv6 addresses (IPv4 mapping) and have been assigned the prefix 0:0:0:0:0:ffff (or, using CIDR notation, ::ffff:0:0/96). For IPv4-mapped addresses, it is possible to use a mixed notation; for example, the IPv6-mapping of the IPv6 address 192.168.0.1 can be expressed as ::ffff:192.168.0.1.

IPv6 also introduces the concept of anycast. Similar to multicast, in anycast many hosts share a single destination address, forming a group of potential receivers; however, contrary to multicast, packets are routed to a single member of the group. Anycast is used to implement load balancing and to increase service reliability. Originally IPv6 was mostly suited for connectionless protocol (UDP); however, implementations now exist that make it suitable for connection-oriented protocols such as TCP.

Despite the great potentialities of IPv6, IPv4 is still the most used network protocol, largely due to the use of alternative solutions based on private IPv4 addresses, NATs, and virtual servers that have all helped to alleviate the critical lack of public addresses. However, the problem of exhaustion of IPv4 addresses returned to the news on February 3, 2011, when the final five blocks of Class A IPv4 addresses (each of them containing about 16.7 million addresses) were officially given out. The transition to IPv6 is really getting closer.

2

DOMAIN NAMING SYSTEM

Hosts are identified in the Internet by means of Internet Protocol (IP) addresses. Despite the use of the dotted decimal notation to try to make IP addresses easier for humans to handle, they are still expressed as collections of numbers that are comparable in length to a traditional telephone number. Although this length is shorter than IPv6 numbers, it still makes it unlikely that human beings will easily remember very many IPv4 addresses.

Certainly a name is much easier to remember than a collection of numbers. Thus, telephone numbers are usually stored in "address books," where users can search by providing a string corresponding to the name of the entity (person or organization) they are seeking.

Looking for a Web or a File Transfer Protocol (FTP) site today is almost always realized using domain names rather than directly looking through lists of IP addresses. Similarly, an email address or a Session Initiation Protocol (SIP) Address of Record usually contains the domain and the name of the server to which they refer rather than an IP address.* Other than a matter of human memory limitations, however, there are other reasons behind not using IP addresses directly.

Consider a collection of services provided by a given server serving many clients. If the IP address assigned to the server changes, then all the clients need to update their records with the new IP address. This operation might be impractical in the context of a local area network (LAN) but is certainly impossible in the open Internet, where the clients are usually unknown to the server. Furthermore, if it is necessary to migrate some of the services to another server (e.g., for scalability rea-

* To use an IP address instead of a domain name in an email address, the address must appear between square brackets, such as example@[10.0.245.254]. Sometimes, however, this may bring unexpected reactions, as not many mail servers support this notation.

sons), then the client will have to add a second IP address to its records, associating this second address to the services that have been migrated.

An even more difficult scenario is where a network administrator, to implement a load-balancing policy, wishes to add a second server that provides the same services as the first one. If the clients refer to the server by using its IP address, then they would need to add a second record with the new IP address and then make a client-side about which server to connect with.

The Domain Name System (DNS) is an abstraction layer decoupling the identifier of a host from its IP address, which allows these, and other, example problems to be solved efficiently. The primary advantages of the DNS are that it is not necessary to update the clients directly and that it is possible to use more easily remembered mnemonic names instead of numbers. The DNS defines a protocol that is essential to the effective functioning of the open Internet; it would be impossible to notify Internet clients when an IP address changes since the clients would frequently be completely unknown to servers.*

A DNS also permits the creation of Virtual Servers, which help to prevent excessive growth of public IP addresses† and is essential to many other applications that will be described in the following.

Domain Names

DNS names (RFC 1034, Mockapetris 1987a) are made of dot-separated strings:

```
string.string.string.….string.
```

* In the early days of the Internet, when hosts were relatively few and well-known, this was actually done just using a single file enumerating host names and their addresses. This file was maintained by the Network Information Center (NIC), which has now evolved into the Internet Assigned Numbers Authority (IANA) and accessed using FTP (RFC 952 and RFC 953).

† Consider the following use case: an organization providing many different services owns just one public IP address, assigned to a host that acts as a front end for all incoming requests. The front-end host could dispatch requests to different internal hosts, each of which has different IP addresses, inside the organization's Local Area Network (LAN). The actual dispatching operation could be performed using the mnemonic address transmitted inside an application layer request (e.g., a Hypertext Transfer Protocol [HTTP] request) rather than the IP address contained in the IP packet.

The first rightmost string is called the top-level domain (TLD), the second rightmost string is called the second level domain, right through to the lowest level domain on the left. There are two different kinds of TLDs: global TLDs and country-code TLDs. Familiar examples of global TLDs are ".com" (corporations and commercial entities), ".net" (networks of computers, also now including those that belong to Internet Service Providers [ISPs]), and ".edu" (educational institutions). Country-code TLD examples are ".us" (United States), ".fr" (France), and ".jp" (Japan). Country-code TLDs use two-letter country codes as defined in ISO 3166.* RFC 2606 (Eastlake and Panitz 1999) reserves four TLDs (.test, .example, .invalid, .localhost) for testing, documentation, examples, and local loopback purposes.

The maximum length of a domain name is 255 characters, whereas each string appearing in each domain level of a DNS name can be up to sixty-three characters in length. These limitations are due to the fact that the DNS protocol uses both User Datagram Protocol (UDP) and Transmission Control Protocol (TCP) connections (both of them on port 53). Common DNS queries are usually performed using UDP datagrams, whereas TCP connections are established only in particular cases (e.g., during zone transfers, which will be described later). The maximum length of a domain name is therefore designed so that a complete name, together with other fields defined by the protocol, could be easily transported in a 512-bytes long UDP datagram. The UDP datagram contains a "truncation bit" that is set whenever the length is greater than 512 bytes (this usually happens in DNS responses). In this case, the DNS client is made aware that the server can satisfy the request and can repeat the request to the server using a TCP connection.

Domain names are case insensitive. The final dot appearing in a domain name is part of the notation and refers to a domain called *root domain* or just *dot*. When the final dot is omitted, the name could be relative to another upper-level domain (assumed it exists), administered by the server in which the name is stored. When the final dot appears, instead, it has the meaning of a full stop; that is, the name is an absolute name ending in (or beginning with) the root domain.

* The ISO 3166 code lists: http://www.iso.org/iso/iso_3166_code_lists.

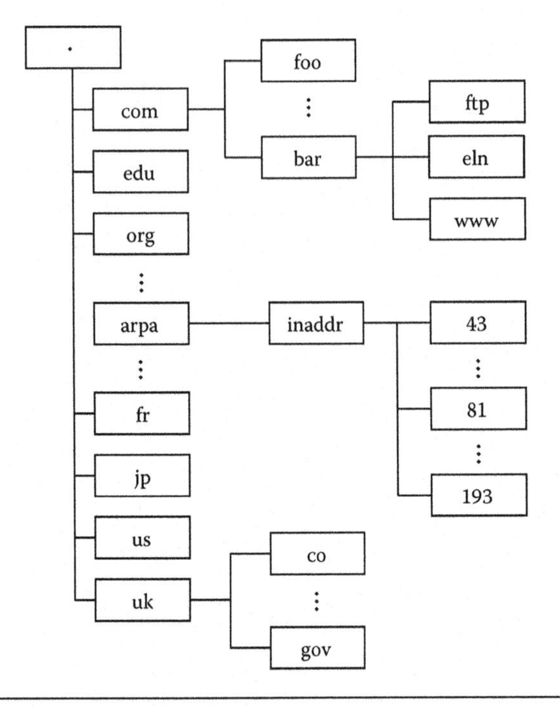

Figure 2.1 Domain names are hierarchical.

The DNS is implemented as a distributed database. The distribution is based on the rationale that an entity (individual or organization) assigned a given domain of any level is completely responsible for managing the names under the subordinate levels of its assigned domain (Figure 2.1). For example, if Example Inc. is assigned the domain *example.com*, it is allowed to define any names behind example.com that it chooses. Thus, it will be responsible for the third-level domain domain1.example.com as well as the fourth-level domain subdomain3.domain2.example.com. As well as being assigned to domains, names are also assigned to hosts. These are simply identified by their assigned name followed by a dot and then their domain name as a suffix. Therefore, the name myhost.domain1.example.com identifies the host *myhost* inside the domain domain1.example.com owned by Example Inc.

Other than translating domain names into addresses, the DNS also performs the reverse operation, which is referred to as *reverse translation*. To efficiently perform this operation, IPv4 addresses are represented as special domains. An IPv4 domain, corresponding to an

IPv4 address such as x.y.w.z, is a subdomain of the TLD inaddr.arpa and assumes the form

```
z.w.y.x.inaddr.arpa.
```

When used in this way, the elements of the IP address are written in the reverse order from that used in the normal IP address dotted notation. Note that this notation allows addresses for hosts as well as for networks to be expressed. For example, the network 195.33.43.0/24 has the corresponding domain 43.33.195.inaddr.arpa.

It is a little more complex to handle the DNS for IPv6 addresses than for IPv4 addresses, but their management models do share similar roots. Special records containing 16 bytes (instead of 4) were used to store IPv6 addresses in DNS databases.* Reverse resolution was performed as for IPv4 addresses using a registered TLD named IP6.INT. Later, a more efficient lookup strategy was introduced (RFC 2874, Crawford and Huitema 2000), together with new record types called A6 and DNAME and a new reverse domain IP6.ARPA. The purpose of the new record types was to simplify multihoming and network renumbering† by exploiting the inherent hierarchical mechanism that was already part of IPv6 specifications and mapping it to the DNS hierarchical domain structure in the most natural way. However, before describing DNS handling of IPv6 addresses, it is necessary to take a closer look at the DNS architecture and its basic procedures.

DNS Architecture

Information on DNS names and their IP addresses together with other kinds of information are stored in name servers (NS). Each

* These records were called "AAAA" records, in order to distinguish them from the ordinary "A" address records used to store IPv4 addresses. "A" records and many other DNS record types will be discussed in the following.

† Multihoming is the ability for a "site" (i.e., a collection of local networks and their nodes) to be reached through different addresses, passing through networks operated by different network providers, thus increasing reliability and fault tolerance. Network renumbering is the facility to adjust network addresses in a seamless fashion, that is, without changing all DNS records describing the nodes (and subnetworks) belonging to the network.

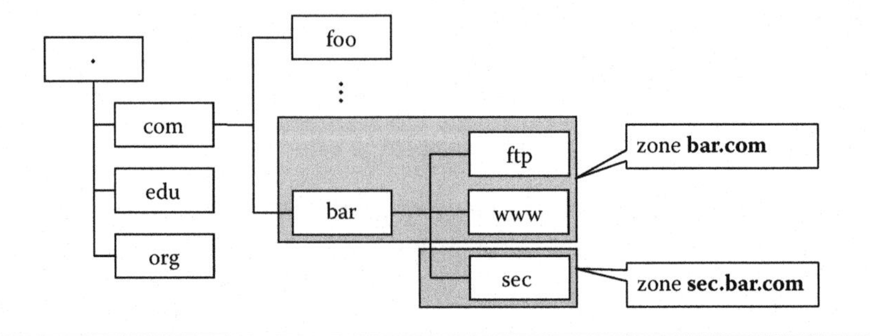

Figure 2.2 Two zones: bar.com and sec.bar.com.

name server could administer an entire domain together with its subdomains or could choose to delegate a subdomain to another name server. The part of a domain administered by a given name server, together with its subdomains not delegated to other name servers, is called a *zone* (Figure 2.2).

There could be different kinds of name servers. A primary name server loads the data about its zone from a database saved on its mass storage memory. A secondary name server acquires these data by copying the data from the primary name server (an operation called *zone transfer*) at regular intervals. The data stored within both the primary and secondary name servers are called *authoritative data.* Other than authoritative data, the primary and secondary name server must also contain nonauthoritative data. These data are usually related either to the root name servers (i.e., the name servers that administer the *root domain,* which are well-known and therefore permanently stored in the local mass storage of a name server) or to cached data obtained from answers to queries that have been answered by other name servers during a process called *iteration,* which will be described in detail later. Name servers that contain only nonauthoritative data and no authoritative data are called caching (only) name servers.

A name server client is called a *resolver* and is usually implemented in the operating system of a computer connecting to the LAN and to the Internet. It is possible to configure the resolver with the name of the local domain and the address of the name servers (the primary server and one or more secondary name servers) to use. It is also possible to provide some preconfigured addresses, usually written into a static file. Whenever an application, such as a web browser

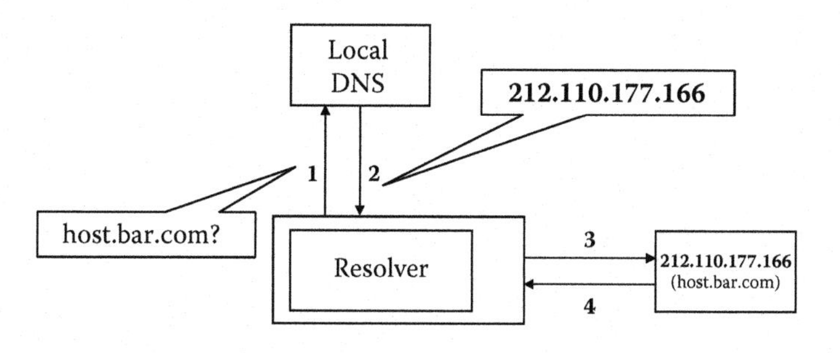

Figure 2.3 A nonauthoritative answer provided by the local name server.

or an FTP client, asks for a domain name resolution, the resolver looks in its local cache. If no answer can be found, it sends a query to the primary or any secondary name server (trying each configured name server in turn if no answer is returned from the previous server within a given amount of time). The contacted name server may be an authority for the requested data; in this case it answers the query directly, marking the answer as an authoritative answer. If it is not an authority, it checks its cache; if the answer is found in the cache, it sends back a response that is, however, marked as a nonauthoritative answer (Figure 2.3).

If no answer is found in the cache, the name server starts the iteration process. The first step consists of contacting the root name server to get the address of the name servers authoritative for the TLD part of the domain name to be resolved. Obviously, the name server may already have this information in its cache, in which case this step is skipped. Then, it interrogates the TLD name server, which provides a response that contains the address of the name server authoritative for the subdomain (this information may also be cached). Finally, the name server turns to the name server authoritative for the subdomain and obtains an authoritative answer, which is transmitted to the resolver and is locally cached for its own use (Figure 2.4). The ancillary queries made to different name servers at different domain levels are called *iterative queries.*

This schema allows for a variant: either the iteration can be performed by the name server contacted by the resolver, or the resolver can delegate to another name server (called *forwarder*) to perform all the operation on its behalf and return a final response. This kind of query is called *recursive* (Figure 2.5), as the client is asking the server

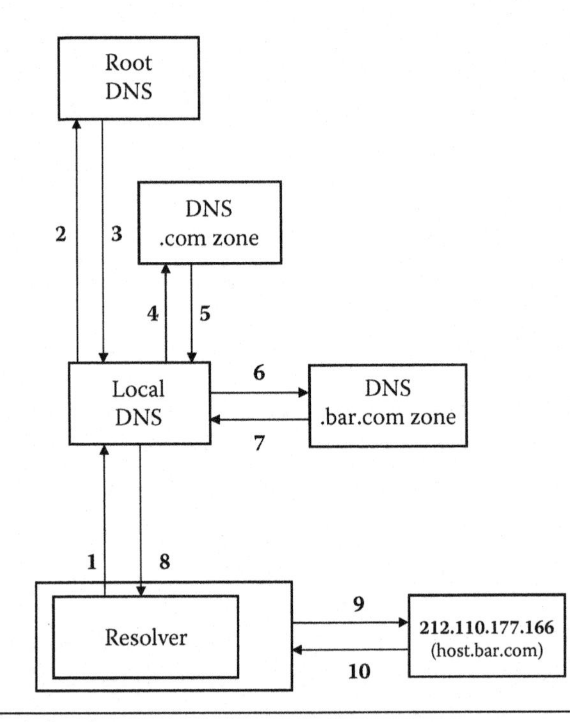

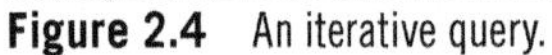
Figure 2.4 An iterative query.

to fully resolve the domain name and provide a final answer. Note that the original query by the resolver to the primary or secondary name server is, by definition, also a recursive query.

Resource Records

Name servers are part of the distributed DNS database and contain DNS data in forms of resource records (RRs). Each RR has a structure composed of six fields (Figure 2.6). The first field is the *name,* followed by the record *type.* The type provides meta-information about the kind of information contained in the record itself. There are different record types, the most common of which are the following:

Type A: A host address. This record describes an IPv4 address associated with the DNS name contained in the *name* field.

Type NS: Defines the DNS name of the authoritative name server for the domain described in the *name* field. This record is used to delegate the administration of a DNS zone to another name server.

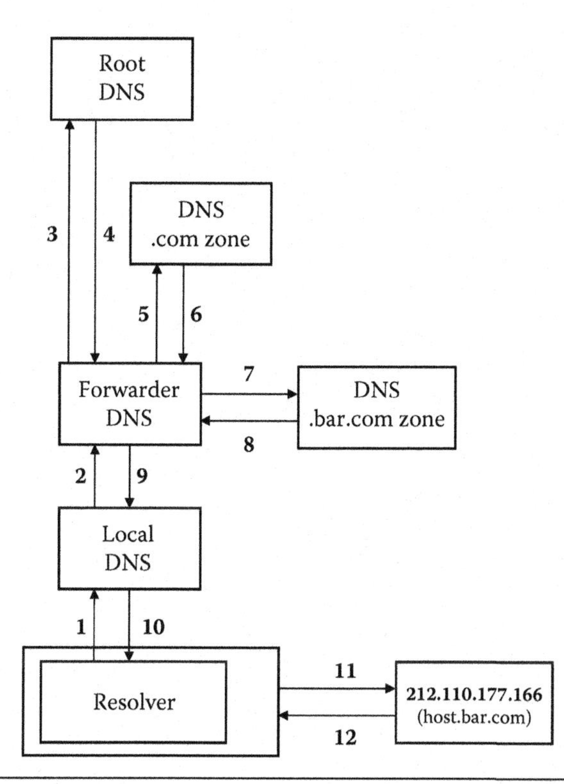

Figure 2.5 A recursive query.

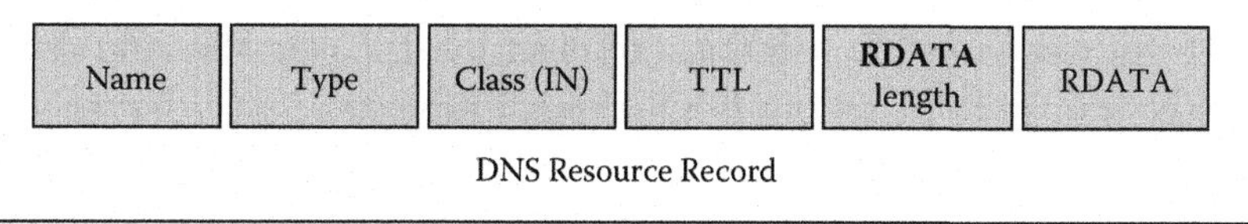

Figure 2.6 A resource record.

Type PTR: Used for reverse translation. The value is a domain name whereas the name field is an address.*

Type CNAME (canonical name): A synonym of the DNS name contained in the first field.

Type SOA (Start of Authority): The record defines the DNS name of the authoritative name server that is authoritative for this particular domain.

* This type was introduced later, after RFC 1035 (Mockapetris 1987b), which still used the term *inverse query*. The inverse query, which is no longer supported by many implementations, did not use PTR records but, less efficiently, looks for the domain name (or domain names) corresponding to an IP address in traditional A records.

The third resource record field is the record *class* that, in case of the Internet, is always set to the abbreviation IN. Other classes (e.g., Chaos, Hesiod) are possible as well but indeed less common. The fourth field is the time to live (TTL), which indicates how long the record can be kept valid in a nonauthoritative server's cache. The last two fields are for the record value. The fifth field defines the length of the value, which is contained in the last field (called RDATA). The value depends on the record type. Most common cases include an IP address for records of type A, a domain name when the record type is NS, CNAME and PTR, and a string in a record of type TXT.

A6 records, which correspond to IPv6 addresses, require further explanation. Compared with A records, an A6 record contains a complex RDATA field composed of three parts. The first part is called *prefix length* and is a decimal number in the range 0–128. The second part corresponds to a part of the IPv6 address, and the third part is a DNS name called the *prefix name.* The word *prefix* refers to a part of the target IPv6 address that precedes the part of the IPv6 address contained in the RDATA field of this record. This information can be found in another A6 record. In fact, a single IPv6 address is divided into several A6 records called *record chains,* reflecting the inherent hierarchical structure of IPv6 addresses. To resolve a DNS name into an IPv6 address using A6 records, a resolver must look at the A6 record that contains the matching DNS name, in the same way that this is done for A records. Unlike A records however, the RDATA field of the A6 record will typically contain only the last part of the IPv6 address and not the whole of it. However, together with this information, the RDATA field will also provide the length of the prefix—say n, with $n \neq 0$—and the (not null) DNS name assigned to the prefix. This means that the IPv6 address contained in the field represents only the rightmost portion of the whole IPv6 address. To find the remaining leftmost n bits, the resolver performs a second step, looking for the name of the prefix, which is, in turn, contained in a further A6 record (the second A6 record of the record chain). In that record, the RDATA field will contain a second portion of the queried IPv6 address, together with a further prefix, and a prefix length, say m. This second portion of the IPv6 address has to be left-concatenated with the first portion found in the previous step, and contributes $128-m$

bits to the whole IPv6 address length. Typically, it will be $m < n$,* meaning that the resolver is on the way to rebuilding the whole IPv6 address, literally bit by bit. The iteration then continues to find the A6 record matching the new prefix, until an A6 record containing no prefix name and a prefix length equal to zero is found. The RDATA field of that record contains the leftmost portion of the IPv6 address that, left-concatenated with the rest of IPv6 address built step by step, will finally form the queried IPv6 address.

It is easy to see that splitting an IPv6 address into several A6 chained records is useful, as it effectively handles the case where entire networks and subnetworks change their address (network renumbering). For example, if the owners of site X homed by a network provider A decided to change provider, say with network provider B, the site would have to change all its IPv6 addresses (corresponding to all its subnetworks and nodes). With the record chain mechanism, it is enough to change a single record in the record chain, the one containing the address prefix corresponding to provider A, replacing its contents with the address prefix of provider B. All IPv6 addresses assigned to the site would then change seamlessly. It is also possible, for site X, to be served by both provider A and provider B (multihoming). In this case, the record containing the address prefix of provider B is added without replacing the original one. Each network (and each node) in site X will be then reachable by two IPv6 addresses, differing only in their prefix.

The reverse resolution procedure is similarly performed in a sequence of steps, using a special bit string notation for IPv6 addresses and introducing a new record type called Delegation Name (DNAME). The bit string notation mandates that an IPv6 address, expressed in classless interdomain routing (CIDR) notation, is written as a string that begins with a backslash, is followed by the *x* character (as in hexadecimal), and then contains the digit sequence of the IP address enclosed in square brackets.

For example,† the IPv6 address

```
2345:C1:CA11:1:1234:5678:9ABC:DEF0
```

* If this is not the case, it means that an error in the resolution process has occurred.

† The following example is taken from RFC 2874 (Crawford and Huitema 2000).

is expressed as

```
\[x234500C1CA110001123456789ABCDEF0/128]
```

The DNAME record type is similar to CNAME, except that it allows an alternative name to be assigned to an entire subtree of the domain name space rather than a single node.

As with IPv4, IPv6 addresses are represented as special domains. An IPv6 domain is a subdomain of the TLD IP6.ARPA, and IPv6 addresses are names inside that domain. For example, the previous address could be thought as a name of the TLD IP6.ARPA domain.

```
\[x234500C1CA110001123456789ABCDEF0/128].IP6.ARPA
```

A resolver trying reverse resolution of this IPv6* address starts by initiating a query for this name to a name server authoritative for IP6. ARPA. The response, however, is likely to be a DNAME record containing the domain name associated with the first part (leftmost bits) of the address, for example,

```
\[x234500/24].IP6.ARPA. DNAME V6.EXAMPLE-TLA.ORG.
```

This record states that the first part of the address (24 bits) is an alias for the domain name V6.EXAMPLE-TLA.ORG. Therefore, the resolver could look for the remaining part of the bit string encoded address under the domain V6.EXAMPLE-TLA.ORG:

```
\[xC1CA110001123456789ABCDEF0/104].V6.EXAMPLE-TLA.ORG
```

To initiate this query, the resolver will typically contact a name server for the zone V6.EXAMPLE-TLA.ORG. This latter in turn could answer stating that the first part (leftmost bits) of the bit string-encoded address is an alias for the domain name IP6.FOO.NET:

```
\[xC/4].V6.EXAMPLE-TLA.ORG. DNAME IP6.FOO.NET.
```

* To better illustrate the procedure, it has been assumed that the consulted servers do not provide recursion and that the resolver does not have cached information.

The resolver would query a name server for the zone IP6.FOO.NET, looking for the remaining part of the bit string encoded address under the domain IP6.FOO.NET:

```
\[x1CA110001123456789ABCDEF0/100].IP6.FOO.NET
```

In this way the iterations continue until the whole address is resolved into a DNS name.

DNS Operations

The most common operation in the DNS protocol is DNS QUERY, which consists of a request and a response. The packet format for both the request and the response is composed of five sections. The first section is the header. The header is composed of the following:

1. A 16-bit DNS query identifier, used to correlate request and response
2. 16 control bits, stating whether the packet represents a request or a response, whether the answer is authoritative or not, whether it is truncated or not, the type of query (operation code*), whether recursion is desired or available from a name server, whether a query interpretation error occurred
3. The number of resource records contained in the four following sections of the packets: (a) queries, (b) answers, (c) authoritative nameservers, and (d) additional records. These sections are all filled with resource records different from resource records stored in DNS servers and may be complete or partially complete and present specific value in some of their fields.

The queries section contains the actual queries (one or more). The resource records in this section contain the three resource record fields *name*, *type*, and *class* already described. Specific additional values for the field *type* include AXFR, IXFR, and *. The first two are used to request a full zone transfer or an incremental zone transfer (which

* The operation code is a constant in packets containing DNS queries; however, since the same header is used in DNS extensions, described later in the text, the code is meaningful as it defines the type of operation that the client is requesting (e.g., query, update, notify).

will be described in the following); the last stands for a request for all records.

The answers section trivially contains the resource record returned by the server as a response to a query. In the case of record not found, the RDATA field in the returned record contains one of the following values (RFC 2306, Parsons and Rafferty 1998):

NXDOMAIN, if there is no domain corresponding to one defined in the query;
NOERROR_NODATA, if the domain exists, but the specific record requested by the query does not;
Server failure, indicating either a problem in the contacted name server, or a problem with the primary name server (reflected in the contacted name server);
Unreachable server.

This information may also be cached in the client.

Following the answers section, the authoritative servers section contains an enumeration of the authoritative name servers for the hosts appearing in the resource records listed in the answers section. Finally, there is an additional information section frequently used to transmit a resource record containing the IP addresses of the authoritative name servers.

An example of DNS query is as follows:

```
DNS query
...
Queries
  #1 host.bar.com: type A, class IN
    Name: host.bar.com
    Type: A (Host Address)
    Class: IN (0x0001)

DNS response
...
Queries
  #1 host.bar.com: type A, class IN
    Name: host.bar.com
    Type: A (Host Address)
    Class: IN (0x0001)
Answers:
  #1 host.bar.com: type A, class IN, addr 212.110.177.166
    Name: host.bar.com
```

```
    Type: A (Host Address)
    Class: IN (0x0001)
    Address: 212.110.177.166
Authoritative Nameservers
  #1 host.bar.com: type NS, class IN, ns alpha.bar.com
    Name: alpha.bar.com
    Type: A (Host Address)
    Class: IN (0x0001)
    Nameserver: alpha.bar.com
  #2 host.bar.com: type NS, class IN, ns beta.bar.com
    Name: beta.bar.com
    Type: A (Host Address)
    Class: IN (0x0001)
    Nameserver: beta.bar.com
Additional Records
  #1 beta.bar.com: type A, class IN, addr 212.110.117.1
    Name: beta.bar.com
    Type: A (Host Address)
    Class: IN (0x0001)
    Address: 212.110.117.1
  #2 alpha.bar.com type A, class IN, addr 212.110.117.2
    Name: alpha.bar.com
    Type: A (Host Address)
    Class: IN (0x0001)
    Address: 212.110.117.2
```

Together with DNS QUERY, other operations have been introduced in the context of the DNS protocol and are referred as DNS extensions. For example, DNS UPDATE, introduced with RFC 3007 (Wellington 2000), allows resource records to be modified within a given zone. Obviously, only primary servers can execute the corresponding update operation; therefore, a DNS UPDATE packet should be sent to primary name servers only (secondary name servers receiving a DNS UPDATE packet may forward it to the primary name server). The update operation is not performed on the fly by the receiving server but is recorded in a journal file and performed at next restart of the server.

A DNS UPDATE packet is composed of five sections, whose structure is similar to those already described for the DNS QUERY packet. The header contains:

1. A 16-bit DNS identifier, used to correlate request and response
2. 16 control bits, stating the type of packet (DNS UPDATE), whether the packet represents a request or a response, and the response error code

3. The number of resource records contained in the four following sections of the packets: (a) zone, (b) prerequisites, (c) update, and (d) additional information. The zone section indicates which zone has to be updated. It contains a single record of type SOA, class IN, which specifies the zone name. The prerequisites section contains a list of conditions that must be fulfilled before executing the update. If one of the conditions fails, then the server must not proceed with the update. These conditions are expressed using the same syntax as resource records description are but interpreting the syntax in the way shown in Table 2.1. Similarly, the update section contains the records that have to be inserted or deleted from the zone. These changes are expressed using the same syntax as resource records description, but interpreting the syntax in the way shown in Table 2.2.

Table 2.1 Interpretation of the Prerequisite Section in a DNS UPDATE Message

PREREQUISITE EXPRESSION	PREREQUISITE INTERPRETATION
NAME and TYPE are given, RDATA length is zero, RDATA is empty, CLASS is ANY, TTL is zero.	The zone contains at least one RR record of a given NAME and TYPE.
NAME is given, RDATA length is zero, RDATA is empty, CLASS is ANY, TYPE is ANY, TTL is zero.	The zone contains at least one record of a given NAME.
A set of records with NAME and TYPE given, RDATA length is zero, RDATA empty, CLASS is specified in the zone section, TTL is zero.	The zone contains a set of records of a given NAME and TYPE.
NAME and TYPE are given, RDATA length is zero, RDATA is empty, CLASS is NONE, TTL is zero.	The zone does not contain any record of a given NAME and TYPE.
NAME is given, RDATA length is zero, RDATA is empty, CLASS is NONE, TYPE is ANY, TTL is zero.	The zone does not contain any RR record of any TYPE with a given NAME.

Table 2.2 Interpretation of the Update Section in a DNS UPDATE Message

UPDATE EXPRESSION	UPDATE INTERPRETATION
NAME, TYPE, TTL, RDATA length, CLASS, and RDATA are given.	Add RR records.
NAME and TYPE given, RDATE length is zero, RDATA is empty, CLASS is ANY, TTL is zero.	Remove a set of RR records of a given type.
NAME is given, TYPE is ANY, CLASS is ANY, RDATA length is zero, RDATA is empty, TTL is zero.	Remove all RR records of a given name.
NAME, TYPE, RDATA length, and RDATA are given, CLASS is NONE, TTL is zero.	Remove one RR record.

One application of DNS UPDATE is the incremental zone transfer (IXFR). The primary name server sends a DNS UPDATE packet containing the SOA record of the zone that has changed. This record in turn contains the serial number associated with the current state of the zone. The serial number is maintained by the primary name server and is updated every time a change in the zone occurs. The secondary name server checks the serial number of its cached SOA record. If this is outdated, it sends a request containing a copy of its SOA record (with the outdated serial number). Finally, the primary server sends the updates, containing the record to be added and deleted. If the serial number communicated by the slave is too old, then the primary name server could reply with a full zone transfer (AXFR) instead of using an incremental zone transfer.

DNS NOTIFY, specified in RFC 1996 (Vixie 1996), allows a primary name server to send a notification to a secondary name server whenever there is a zone change.

A DNS NOTIFY message is triggered by a change in the SOA record contained in the primary name server. Upon receiving the DNS NOTIFY message, the secondary name server should query the primary to obtain the corresponding SOA record. If this record contains a serial number greater than the one contained in its copy of this SOA record, then the secondary server should ask for a zone transfer.

The packet format of a DNS NOTIFY message is the same as the one defined for a traditional DNS query (with the operation code properly set to NOTIFY). In some cases, the primary server can send an indication of the change directly in the notification message, specifying the name, class, type, and value of the changed records in the queries section of the DNS NOTIFY packet. However, these records are intended not as updates to be performed directly by the secondary name server but just as indications of the changes that are going to occur when the secondary name server asks for the zone transfer.

3
Hypertext Transfer Protocol

Hypertext Transfer Protocol (HTTP; Fielding et al. 1999) is indeed the most popular protocol on the Internet, which leads many people to think that HTTP and the web are the same as the Internet. Actually, HTTP is now replacing many protocols that have decreased in popularity. Many of these protocols have even disappeared since the invention of HTTP gateways that perform conversions between HTTP and the target protocols, allowing information to be accessed using popular web browsers instead of protocol-specific clients.

Brief History of HTTP

Compared with other Internet protocols, HTTP is quite young. What is nowadays widely known as the World Wide Web actually consists of the winning combination of a transfer protocol (HTTP) and a presentation language (Hypertext Markup Language, HTML). The web was introduced in 1989 by Tim Berners-Lee at the European Organization for Nuclear Research (CERN) to improve information access within the physics community. Previously, researchers who wanted to retrieve information over the Internet mostly relied on the File Transfer Protocol (FTP; Postel and Reynolds 1985), which has been in place since the early 1970s. Compared with the modern web, FTP appears as very primitive. It does not provide any lookup facility and requires that a user looking for information should know in advance the address of the resource or resources containing the desired information.

Before HTTP, Archie (Emtage and Deutsch 1992), WAIS (Kahle and Medlar 1991), Gopher (Anklesaria et al. 1993), and other protocols—all of them file oriented—were used to try to solve this problem. Archie was a simple but useful program that was able to anonymously access FTP servers and record the names of the files they contained, creating a searchable global catalog of file names available for download. A client could then simply perform a query on the catalog to retrieve the local FTP site where files with names that matched the input expression resided. The introduction of WAIS extended the indexing capabilities, allowing searches to be made inside file content as well as across file names. Using WAIS, Gopher further extended this capability, allowing information to be arranged in a hierarchical fashion through the use of menu items. Gopher servers were also distributed and the user could seamlessly pass from one Gopher server to another by navigating menu entries. This was a particularly innovative feature in the early 1990s. The introduction of a similar feature for hypertext links, which existed prior to 1989 but were confined within the boundaries of a single computer, was the key to success of HTTP and paved the way for the extraordinary development of the Internet over the last twenty years.

Despite its origin, today HTTP is much more than a way to retrieve hypertext or multimedia files in the Internet. The Representational State Transfer (REST) is the architectural style built on HTTP and best highlights how extensively the protocol could be used for sharing and controlling the state of distributed data resources over the network.

Standards and technologies that rely on the facilities offered by HTTP are numerous. Even novel proposals for new content-centric networks, which replace the whole network and transport layers, can still be backward compatible with existing HTTP browsers and software, leading to simplifications in their design (as content-centric networks may natively implement some content management facilities that are in HTTP specifications). Additionally, the well-known aspects of HTTP make it an attractive candidate protocol to use. As previously described, due to the lack of Internet Protocol, version 4 (IPv4), addresses, network nodes in many organizations are behind a Network Address Translation (NAT), and for security reasons, many organizations' firewalls allow HTTP messages to pass through only their ports. There is a common perception that a web browser can be considered to be a secure sandbox, thus reducing or removing the need to install a separate new executable sandbox application. Mobile developers are also experimenting with similar approaches based on enhancing security by minimizing access by means other than HTTP. For example,

Java Micro Edition* specifications mandate only Java application programming interface (API) for HTTP connection handling, with the API to manage Transmission Control Protocol (TCP) and User Datagram Protocol (UDP) connections being optional. Even when it is possible to use UDP on a mobile device, UDP could be blocked by network operators or not allowed by firewalls. In some cases, only web browsing is permitted, sometimes with access allowed only via a proxy server. Considering all of these diverse applications of HTTP, it can be seen that there has been a recent explosion of attempts to overload the original nature of the protocol itself so that its usage has crossed the boundary of a request–response protocol to now become employed as a streaming or even a real-time protocol. Some of these protocols will be described at the end of this section, after the internals of the HTTP protocol have been analyzed in more details.

Despite all this, HTTP current specifications still present some limitations. At the time of writing, the HTTPbis Working Group at the Internet Engineering Task Force (IETF)† is addressing possible fixes to ambiguities and lack of details in the present standard. Specifications for the new, more efficient HTTP/2.0 are also expected from this group.

URL, URN, URI, and IRI

The web uses global identifiers to allow a client to find and access a specific resource. Historically two different schemas have been created, the Uniform Resource Name (URN; RFC 1737 [Sollins and Masinter 1994] and RFC 2141 [Moats 1997]) and the Uniform Resource Locator (URL; RFC 1738 [Berners-Lee, Masinter, and McCahill 1994] and RFC 1808 [Fielding 1995]). The aim of a URN is to provide persistent and immutable names for a resource, independent of where the resource is located. In contrast, a URL provides a (not necessarily persistent) address where the resource can be accessed.

The dualism between "identifiers" and "locators" has been the heart of several debates on the web (Hayes and Halpin 2008). Today the trend is in favor of using URLs, because of their ability to be resolvable without requiring specialized software (to retrieve a resource it

* Java Micro Edition is a Java-based programming language for handheld devices.

† IETF HTTPbis Working Group at IETF: http://datatracker.ietf.org/wg/httpbis/charter/.

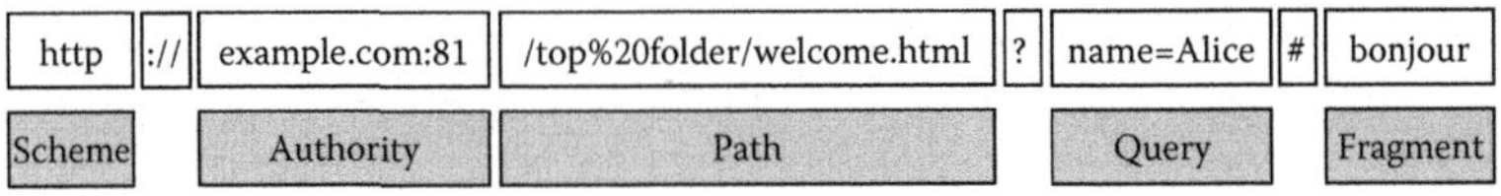

http	://	example.com:81	/top%20folder/welcome.html	?	name=Alice	#	bonjour
Scheme		Authority	Path		Query		Fragment

Figure 3.1 A URIRef (HTTP schema).

is enough to put its URL into a web browser bar), and makes them persistent as much as possible (Berners-Lee 1998).

However, the two schemas have been conceptually unified under the notion of Uniform Resource Identifier (URI; RFC 2396 and RFC 3986 [Berners-Lee, Fielding, and Masinter 1998, 2005]), and in practice it is common to use the term HTTP URI to actually refer to a HTTP URL. A URI is composed of at least two components, a *scheme* and an *authority*. The scheme component is case insensitive; however, it is common convention to use lowercase letters. The authority component is scheme dependent and may be case insensitive (noticeably in HTTP, where the authority is a domain name).

A URI may also include a *path* component, a *query* component, and a *fragment* component; in general all of these are case sensitive. Where a fragment component is present, the term URI Reference (URIRef) is usually used instead of URI. Figure 3.1 illustrates a URIRef of the HTTP schema (note that "www" is part of the authority—it is just a host name in the domain example.com—and by no means part of the schema).

Unreserved characters that may be used in minting a URI are chosen from a subset of the US-ASCII set and include letters, digits, dot, dash, underscore, and tilde. Other characters are reserved, but they can appear in a URI as percent encoded, that is, as a triplet made of a percent sign followed by two hexadecimal digits corresponding to the code assigned to that character in the US-ASCII character set. For example, space (US-ASCII character #32) can appear inside the path component of an URI as %20.

With the internationalization of the web, the choice of supporting only Latin characters appeared too limited. To overcome this limitation, a new scheme, the International Resource Identifier (IRI) scheme (RFC 3987, Duerst and Suignard 2005), was proposed. The IRI scheme also supports characters from the Universal Character Set (Unicode Consortium 2012). But apart from this, the syntax, composition rules, and reserved characters are the same as in URIs. As a consequence all valid URIs are also valid IRIs.

Similar to URIs, IRIs can be used to both identify resources and locate resources. For the second purpose, however, IRIs do not define a separated resolution mechanism. Instead, they are mapped into corresponding URIs using percent encoding as following: the UCS non–US-ASCII character is encoded into a sequence of US-ASCII characters using the UTF-8 encoding (RFC 3629, Yergeau 2003), and all of them are then percent encoded. For example,

```
http://example.org/resource/פירנצה
```

is mapped into the URI

```
http://example.org/resource/%D7%A4%D7%99%D7%A8%D7%A0%D7%A6%D7%94
```

Using this transformation, all the existing software able to understand URIs can be reused without change to support IRIs.

HTTP Methods

As a connection-oriented protocol, HTTP is based on TCP. Originally different pairs of HTTP request–response were carried over different TCP connections, which were then finalized when each response was received. A browser, after resolving the host address included in the requested URI, established a TCP connection with the origin server (i.e., the server hosting the wanted resource) using port 80 (or another port specified in the request URI), issued a HTTP command, received a response, and closed the TCP connection. While this mechanism could be considered inefficient but tolerated when the users' Internet connection bandwidth was limited by 56 Kbps modems and the hypertexts contained much more text than pictures, it soon appeared completely unacceptable. This led to a radical change in HTTP 1.1, which introduced persistent connections. A persistent connection allows multiple HTTP request–response pairs to be transported on the same TCP connection (imposing, however, a limit of two simultaneous TCP connections per origin server to prevent possible denial-of-service attacks). In addition, requests are pipelined, meaning that a client may transmit multiple requests in a series over a single connection without waiting for a response. Persistent connections either can be explicitly closed by the client or the server (using the Connection

Table 3.1 General Header Fields

HEADER	EXPLANATION
Cache-Control	Specifies directives that MUST be obeyed by any cache along the request–response path
Connection	Allows the sender to specify options that are desired for a particular connection
Date	The date and time at which the message is originated
Pragma	Implementation-specific directives that might apply to any recipient (including proxies, caches, and gateways) along the request–response path
Trailer	A given set of header fields is present in the trailer of a message encoded with chunked transfer coding
Transfer-Encoding	Indicates what type of transformation (e.g., chunking) has been applied to the message body in order to safely transfer it between the sender and the recipient
Upgrade	Allows the client to specify an additional communication protocol it would like to use if the server supports it (e.g., WebSocket)
Via	Used by gateways and proxies to indicate the intermediate protocols and recipients between the sender and the final recipient
Warning	Used to carry additional information about the status of a message

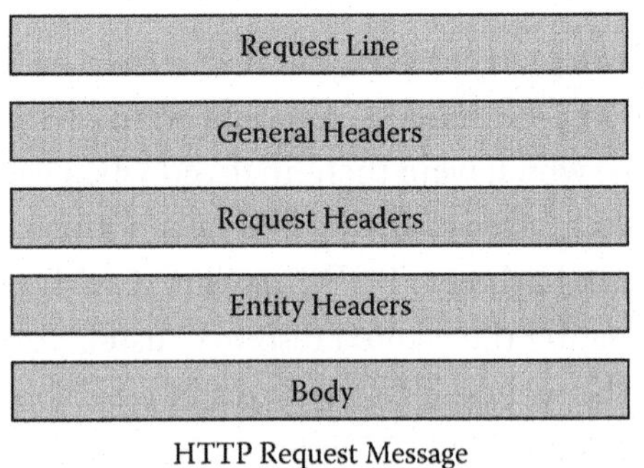

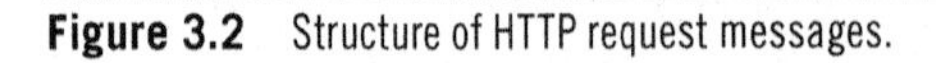
Figure 3.2 Structure of HTTP request messages.

general header specifying the value "close" in a request or a response message; Table 3.1) or are closed after the expiration of a timeout in the server.

A HTTP request (Figure 3.2) consists of one or more lines of text. The first line, called the *request line,* contains the command to be issued to the origin server. The request line has three parts: the method to be applied to the resource, the request URI, and the protocol version in use. This is the minimum mandatory set of information a HTTP request may contain (in the previous version of the protocol, HTTP 1.0, the protocol version was optional).

Table 3.2 Request Header Fields in a HTTP Request Message

HEADER	EXPLANATION
Accept Accept-Charset Accept-Encoding Accept-Language	Specifies acceptable media types, character sets, encoding, and language for the entity contained in the response
Authorization	Used to convey the credentials containing the authentication information of the sender
Expect	Indicates that particular server behaviors are required by the sender
From	If given, SHOULD contain an Internet email address for the human user who controls the requesting user agent
Host	Specifies the Internet host and port number of the resource being requested
If-Match If-None-Match If-Modified-Since If-Unmodified-Since If-Range	Headers used in conditional requests
Max-Forwards	Limits the number of proxies or gateways that can forward the request to the next server in the chain
Proxy-Authorization	Used to convey the credentials containing the authentication information of the sender. Unlike *Authorization* it is consumed by the first proxy that was expecting to receive credentials
Range	Used to request one or more subranges (in terms of bytes) of the entity instead of the whole entity
Referrer	Represents the address (URI) of the resource from which the request-URI was obtained
TE	Indicates what extension transfer codings a client is willing to accept in the response and whether it is willing to accept trailer fields in a chunked transfer coding
User-Agent	Contains information about the user agent originating the request

The request line may be followed by one or more general headers, request headers, or entity headers. General header fields (Table 3.1) apply to both request and response messages and refer to the message being transmitted. The request header fields (Table 3.2) allow the client to pass additional information about the request, and about the client itself, to the server. The entity header fields (Table 3.3) define meta-information either about the resource identified by the request or about the entity contained in the body of the request. A request is terminated by a blank line (Carriage Return Line Feed [CRLF]).

Although HTTP 1.1 defines many methods; only two are mandatory for a general-purpose server. Custom implementations may

Table 3.3 Entity Header Fields in a HTTP Request or Response Message

HEADER	EXPLANATION
Allow	Lists the set of methods supported by the resource identified by the Request-URI
Content-Encoding	Indicates what additional content encodings have been applied to the entity body
Content-Language	Describes the natural languages of the intended audience for the enclosed entity
Content-Length Content-Location Content-MD5 Content-Range Content-Type	Used to indicate the size of the entity body, the resource location, an MD5 digest of the entity, which part is transmitted, its media type
Expires	Defines the date and time after which the response is considered stale
Last-Modified	Indicates the date and time at which the origin server believes the variant was last modified

extend the set of already defined methods. The two mandatory methods are GET and HEAD (all methods are written in upper case and are case sensitive). These methods do not take any action other than retrieval and are defined as *safe methods,* meaning that the intended action requested by the client does not produce side effects. (While it is still possible that particular server side implementations produce side effects, this would happen "behind the scenes" and the client would be not aware, nor would it be supposed to be aware, of them.) The difference between the two methods is that, while the response to a GET method actually carries contents in the response body, there is no content retrieval in response to a HEAD request; only meta-information contained in the header of the response is returned. Clearly this choice improves performances as it avoids transfer of large amounts of data while still allowing the status of the target resource to be inspected. Usage of HEAD includes testing hyperlink validity, testing accessibility, and testing modifications.

Looking at the meta-information that could be retrieved in the response should aid understanding of this facility. The response message (Figure 3.3) is similar to the request message; the request line is replaced by a status line, which contains three elements: the HTTP protocol version, the status code, and an associated textual explanation of the code. The status code is a three-digit code, with the first

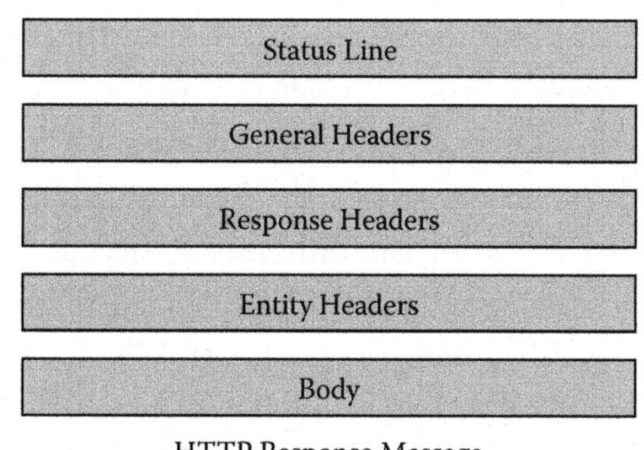

Figure 3.3 Structure of HTTP response messages.

Table 3.4 Response Header Fields in a HTTP Response Message

HEADER	EXPLANATION
Accept-Ranges	Allows the server to indicate its acceptance of range requests for a resource
Age	Conveys the sender's estimate of the amount of time since the response (or its revalidation) was generated at the origin server
ETag	Provides the current value of the entity tag for the requested variant
Location	Redirects the recipient to a location other than the Request-URI for completion of the request or identification of a new resource
Proxy-Authenticate	Used in 407 (Proxy Authentication Required) response messages. The field value consists of a challenge that indicates the authentication scheme and parameters applicable to the proxy for this Request-URI
Retry-After	Can be used with a 503 (Service Unavailable) response to indicate how long the service is expected to be unavailable
Server	Contains information about the software used by the origin server to handle the request
Vary	Determines whether a cache is permitted to use the response to reply to a subsequent request without revalidation
WWW-Authenticate	Used in 401 (Unauthorized) response messages. The field value consists of at least one challenge that indicates the authentication schemes and parameters applicable to the Request-URI

digit indicating the class of error and the other two identifying the specific error (this will be described more fully later).

The response header fields (Table 3.4) are of more interest for the present discussion. Among other useful meta-information, the response header may contain the entity tag (ETag response header field) of the resource, which is a unique identifier used to distinguish between different "versions;" more properly called *variants* of the same resource, that is, resources corresponding to the request URI. It may also include the last-modified response header field, that is, the date and time at which the origin server believes the variant was last modified.

A client that has a cached resource variant may use the entity tag or the age field from a response header to decide whether the resource is out of date. But this is not the only possibility, as HTTP requests allow preconditions to be included in request headers that use entity tags or date stamps. These include the If-Match, If-None-Match, If-Modified-Since, If-Unmodified-Since, or If-Range request header fields (Table 3.2). All these have intuitive meaning. The first two are used to test the current resource's ETag against one or more ETags provided in the request header field. The second two use time stamps to perform the same task. The last precondition requires more explanation. A client that uses the If-Range request header field can choose not to retrieve the complete content, but only the part of it corresponding to a given range (usually in terms of octets composing the content). Combining this feature with the option to validate the resource variant against a specific Etag or time stamps requires two requests: one to check for any modification and the second to retrieve the part of the content fitting the wanted range. However, If-Range allows a shortcut, instructing the server to send the requested part of the content if it is unchanged (validating either the ETag or Last-Modified field) or to send the whole content if there have been changes.

Before considering other methods with side effects, two more safe methods are described. To test which methods can be invoked on a target resource, the OPTIONS method is used; the response to this method includes an Allow header field that enumerates the methods allowed on the resource. If proxy servers are interposed between the client and the origin server, the Max-Forwards request header field may be used in an OPTIONS request to target a specific proxy in the request chain. Similarly, the TRACE method is used to test the path between the client and the origin server or an intermediate proxy (depending on the value of the Max-Forwards field). The target server reflects the received message back to the client, and the content of the Via general-header field is then inspected to check the path; its contents include information on the host, the port, and the protocol used by the proxy or gateway on the chain that handled the message. Hosts within firewall regions are usually replaced with pseudonyms so that the internal network topology is hidden.

A method is said to be *idempotent* if and only if the side effects (if any) of a sequence of one or more method invocations are the same as for a single request. Safe methods are also idempotent.

GET and HEAD are clearly idempotent (as well as safe); OPTIONS and TRACE are idempotent too (having no side effects). Both the DELETE and PUT HTTP methods also share the property of idempotence. Their meaning is intuitive; however, they are still worthy of further consideration. Whenever a client invokes DELETE, on success, the server may respond with one of the three error codes: 200 OK, 202 Accepted, or 204 No Content. The difference between the first and second codes is that the delete action has been enacted in the first case, whereas it may be delayed in the second. The third code is used whenever the action has been enacted, but the response does not carry any content (e.g., no human readable description in the body or similar content). However, even if a server responds with a successful response, it does not mean that the target resource has actually been removed from the origin server. Usually servers implement this command by moving the resource to another, possibly inaccessible, location (a sort of waste bin) rather than removing it.

The PUT method is an example of a command that encloses content (an *entity body*) in the request message. That entity body is intended to be a new variant of the resource corresponding to the request URI, or if no resource was previously assigned to that URI, the new resource that will be pointed to by that URI. A PUT request may include an Allow header field to suggest the methods to be supported by the new or modified resource.

Similarly to PUT messages, POST request messages also carry contents in their body; however, the POST method has to be interpreted differently. While the entity body of a PUT message is intended to be either a new resource variant or the new resource that will be pointed to by the request URI, in a POST message the request URI points to at the resource that will handle the entity body, respecting the intended semantics of the method, that is, "appending." This method may or may not result in the creation of a new resource. In the first case, a 201 Created response message is sent back to the client containing a Location response header with the URI of the newly created resource. In the case when no new resource has been created (because, for example, the posted data represented an SQL statement to be executed by a backend database daemon), a simple 200 OK or 204 No Content message may be generated as a response to the client.

Chunks and Cookies

Origin servers may indicate the size of a response message using the Content-Length header field. For dynamic content, however, the size is often unknown; therefore, the server could include the Transfer-Encoding general-header field (Table 3.1), specifying the value "chunked" to indicate that the response message is transmitted in several chunks.* Conventionally, a zero-length chunk is used to mark the end of the transmitted response message.

Cookies are a means to partially amend the stateless nature of the HTTP protocol. A cookie is simply an arbitrary string transmitted by the server to the client in a response message using the Set-cookie header field. Upon receiving the message, the client stores the cookie locally and includes it in following request messages using the Cookie header field. The inclusion of the cookie in the requests allows the server to recognize the client on subsequent requests.

Representational State Transfer Architectural Style

The REST is an architectural style theorized and first implemented by Roi Fielding (2000). REST replaces the multiplicity of communication protocols in favor of a simple, uniform interface based on interactions by pure HTTP methods.

In line with the original spirit of HTTP specifications, REST defines three kinds of components:

- The user agent, that is, the client that initiates the request
- The origin server, which receives and handles the request
- The intermediaries (proxies or gateways), which perform functions related to security, performance enhancement, or tunneling to other protocols

* With HTTP 1.0, which did not allow persistent connection, the mechanism was even easier, as the server simply closed the connection at the end of the transmission.

The client may (or may not) choose a proxy to which it connects. The gateway, if present, is imposed by the network and automatically reached by the client; in fact, if a gateway is set up, normally any DNS resolver returns the gateway's IP address instead of the origin server's address.

Components communicate by transferring variants or *representations* of resources whose state may vary over time, usually as a consequence of previous REST interactions. HTTP methods are used, and their semantics must be honored (e.g., it is not compliant with a REST architecture using a HTTP GET request to modify a resource, method PUT should be used instead).

It is important to stress that, while the URI in a HTTP request identifies a resource, the returned response contains a representation of it. The representation may not even exist at the time the request is made; it could be created immediately after the request reaches the origin server. This feature allows a sort of "late binding," permitting a server to create on the fly various representations of the same resource according to the client's wishes. For instance, the format of a representation, that is, its media type (RFC 1521, Borenstein and Freed 1993), may be negotiated by clients, based on their capabilities. This is done using the Accept request header field.

To simplify the handling of the requests, each request is stateless. Any session state must be maintained by the client. However, to improve efficiency and performance, some responses—usually, all responses to GET requests—may be cached. Using a HTTP conditional request to the origin server, a cache could obtain information to determine whether each of its cached objects is up to date or needs to be refetched before serving it to the client.

Caches may be shared by multiple clients. Using a shared web cache, an origin server may share some responses with clients other than the one that originally requested them. This technique avoids the extra overhead of recreating a representation for each request. A good design principle to implement a shared web cache is to introduce an intermediary (a *reverse proxy*) between the client and the origin servers. The reverse proxy is the client's primary point of contact and maintains a shared cache with already issued responses from the servers. The client contacts the reverse proxy, which may answer with a cached response; if the wanted response is not found in the cache

Table 3.5 Cache-Control Directives

CACHE-CONTROL HEADER VALUE	SPECIFIED BY	EXPLANATION
no-cache	client/server	Do not return a cached object (client) or do not cache the object in the response (server)
no-store	client/server	Do not store any portion of this message in the cache (e.g., because of private content)
max-age = n s-maxage	client/server	Similar to the Expires header (Table 3.4). Accepts responses whose age is less than n seconds (client) or instructs the cache to revalidate an object after n seconds since its last validation (server). If s-maxage is used, then it is applied only to shared caches (not private caches)
min-fresh = n	client	The client accepts responses that will still be fresh for at least n seconds
max-stale = n	client	The maximum age of an acceptable stale object
no-transform	client/server	Do not compress or modify the object before caching it (server) and do not return such object (client)
only-if-cached	client	Returns only objects which are already in the cache (do not ask the origin server)
public	server	The cache can store this object
private	server	This object contains private data (intended to be returned to a single user) and should not be stored in a shared cache
must-revalidate	server	The cache cannot return a stale object to the client; it revalidates (checks with the origin server) the freshness of the object
proxy-revalidate	server	Revalidates content every time the client requests it. Uses a conditional GET to the origin server to check the current credentials

or is too stale, the reverse proxy asks the origin servers. In this network architecture, the origin servers perform the sole task of creating responses, whereas the reverse proxy takes care of distributing them. Other kinds of shared caches and their possible combinations will be discussed later when describing content delivery networks.

Both the client and the server can specify caching directives including the general-header field Cache-Control in their message (Table 3.1 and Table 3.5). Shared web caches may also host private content, but with special care. The Cache-Control: proxy-revalidate directive is used for such content. Using this directive, an origin server tells the cache to revalidate a content every time the client requests it. Instead of fetching the whole content, however, the cache, upon receiving the request, issues a conditional GET to the origin server. The conditional GET contains the Authorization header specified by the client. If the credentials are valid, then the origin server responds with a 304 Not Modified message and the cache serves the cached content to the client. Otherwise, the origin server responds with a 401 Authorization Required message that forces the client to enter valid authorization information.

Non-REST HTTP-Based Protocols

Recently, standard proposals have been raised to allow bidirectional communication channels over HTTP. An IETF draft (Hickson 2010) describes the WebSocket protocol as a common solution for browser-based applications requiring a full duplex communication channel between the browser and the server (chat, gaming, collaboration, remote controls, real-time applications in general).

WebSocket is actually an upgrade of HTTP and uses its same URI addressing mechanism, except for the two new schemas introduced, ws:// and wss://, respectively, for unencrypted and encrypted connections and ports (80 and 443). Its handshake begins with a particular HTTP request that instructs the server to upgrade the connection to WebSocket. If the handshake is successful, each party can send data independently from the other. A framing technique is used to facilitate event handling.

Bidirectional-streams Over Synchronous HTTP (BOSH; described in Paterson et al. 2010) is a different attempt to emulate a bidirectional stream using multiple HTTP request–responses. Originally developed by the Jabber community as a transport mechanism for Extensible Messaging and Presence Protocol (XMPP; RFC 3920, Saint-Andre 2004), BOSH imposes constraints on the payload of HTTP requests and responses, which should conform to a given Extensible Markup Language (XML) schema. Long-lasting HTTP polling (so-called long polling) and multiple requests (but no more than two requests per time per given web server address as stated by RFC 2616, Fielding et al. 1999) are combined to achieve low latency and cope with disconnections.

Bayeux (Russell et al. 2007) defines a higher level message passing protocol for communications between a browser and a server.

Messages are routed via named channels. The channel is specified in every message and indicates the destination of the message (in the case of a request) or the source of the message (in the case of a response). When Bayeux is transported over HTTP, two simultaneous HTTP connections are exploited to achieve bidirectional communication. Notifications are delivered using either the aforementioned long polling mechanism or a second mechanism called *HTTP streaming*, which consists of a single "endless" HTTP response that is able to carry notifications to the client.

HTTP Authentication Methods

HTTP provides two different types of user authentications: basic authentication and digest authentication (RFC 2671, Vixie 1999). Authentications are often used to protect access from unauthorized users, as in the case of a proxy server in a corporate network allowing only authorized users to surf the web.

With the basic authentication, upon receiving a HTTP request, the server returns a 401 Unauthorized response message including a "challenge" in the WWW-Authenticate response-header field (Table 3.5). The client, usually a web browser, then prompts the user for a username and password, which are transmitted in the Authorization request-header field (Table 3.3) of the next request. Provided the credentials are valid, the server can then respond with a normal HTTP response (otherwise it will resend a 401 Unauthorized message). As the protocol is stateless, the client will have to include the Authorization header field in any subsequent requests to the server.

Using HTTP digest authentication (Franks et al. 1997, 1999), the client can prove that it knows the correct credentials without having to send them in clear to the server. Using its credentials and a random one-time bit string (aka *nonce*) specified by the server in the WWW-Authenticate response-header field, the client uses an application of the MD5 cryptographic hashing to compute a value that is then transmitted to the server. The server does the same computation and authenticates the client if it obtains the same value. The cryptographic hashing provided by the MD5 function should be irreversible, thus preventing an eavesdropper from obtaining the credentials.

SSL/TLS, X.509, and HTTPS

With HTTP authentication, user credentials are transmitted in clear text and are thus subject to eavesdropping. To protect the HTTP connection, more sophisticated cryptographic algorithms are needed. Secure Socket Layer (SSL) is a proprietary protocol to secure TCP/IP communications. Transport Layer Security (TLS; Dierks and Allen 1999; Dierks and Rescorla 2006) is the successor to the SSL and has been standardized by the IETF. TLS may be backward-compatible with SSL (weakening some security features), and therefore it is common to refer to the two protocols with the acronym SSL/TLS. SSL/TLS is application independent, sits on top of the TCP, and seamlessly provides applications with the transport-like interfaces as if it were a common transport protocol such as TCP. However, unlike traditional transport protocols, SSL/TLS adds security features to prevent eavesdropping, tampering, and message forgery.

In summary, SSL/TLS works in three steps. When the communication is established, SSL/TLS authenticates (one or both of) the parties using public key cryptography; then it enables the parties to negotiate a symmetric encryption algorithm and corresponding encryption keys. Finally, the protocol allows the parties to exchange encrypted data.

Although the protocol defines a number of different cryptographic techniques, formats, and protocols, in practice, X.509 certificates and related methods are the usual choice. With X.509, each public key of a private–public key pair is contained in a certificate, that is, a specially formatted document signed by a trusted third party (*certification authority*) that can ensure that the public key is rightfully associated to a given name and assigned use. Scalability is achieved in a hierarchical fashion: the public key of the signer can be verified using a certificate of a superior instance, until so-called root certificates are reached. Genuine root certificates are typically available from private companies that adopt the role of topmost certification authorities.

A certificate is trusted if it is signed by an authority whose certificate is in turn trusted, unless the certificate is found to be in a certificate revocation list (CRL). Signed CRLs are maintained online by certificate authorities to invalidate certificates for which the corresponding private key has been lost.

In a traditional scenario, such as HTTP Secure (HTTPS), the party to be authenticated is usually the server, and the client owns or accesses a certificate storage, which it uses to assess the server's identity. The client software is usually preloaded with a number of root certificates corresponding to the well-known certification authorities; in addition, it usually allows new certificates to be added. However, the same technology could be used to assess a client's identity. In this case, it is the client agent that presents its certificate to the server.

HTTPS was originally introduced in 1994 for Netscape Navigator and used the SSL protocol. The current version of HTTPS, which uses TLS, is specified by RFC 2818 (Rescorla 2000). Except for the fact that HTTPS URLs begin with "https://" and use port 443 by default, HTTPS has identical syntax to the standard HTTP scheme. Using TLS and X.509 certificates, HTTPS provides authentication of a web site, protecting the client from man-in-the-middle attacks. web browsers are released with preloaded certificates that enable seamless use of secure web sites that have brought a public key signed by one of many commercial certification authorities. Revocation is managed using a simple protocol called Online Certificate Status Protocol (OCSP). With OCSP, before establishing a HTTPS connection, the browser sends the HTTPS site certificate's serial number to the certificate authority (or to a delegate) to inquire whether the certificate is still valid or has been revoked.

To protect against eavesdropping and tampering with or forging the contents of the communication, HTTPS and other SSL/TLS-based protocols also provide bidirectional encryption of communications between the client and the web server. In particular, the request URL, query parameters, headers, and cookies are encrypted, but not the host addresses and port numbers, which are needed by the underlying routing and transport.

SECTION II
MANAGING XML DATA

The Extensible Markup Language (XML) is a pillar of data management. Its use has skyrocketed in recent years. An enormous amount of information is stored in XML. Rather than being arranged into tables and columns, in XML data are contained on collections of hierarchical documents that can be stored in file systems or databases.

XML by itself is not enough for managing distributed data, which involves interacting with remote services and transmitting data on the wire. On the Internet and in next-generation networks (NGN) this is realized in a variety of ways. Some of them, such as Web Services, the XML Configuration Access Protocol, and the recent Open Data Protocol (OData) are standard protocols or part of ongoing standardization efforts.

Web Services specifications were originally developed and are still maintained by the World Wide Web Consortium (W3C) and the Organization for the Advancement of Structured Information Standards (OASIS). They are based on a protocol stack on top of the transport layer. The service interface is described using the Web Services Description Language (WSDL), and messages between services are exchanged using an XML-based serialization format, the Simple Object Access Protocol (SOAP).

As opposed to the full-fledged Web Service architecture, the OData is based on the Representational State Transfer (REST) architecture and defines conventions for handling data on the web using the Hypertext Transfer Protocol (HTTP). OData supports different serialization formats including XML and JavaScript Object Notation

(JSON), which are used to encode representations of handled resources. HTTP methods, messages, and headers provide instead most parts of the operational semantics of the protocol.

The XML Configuration Access Protocol (XCAP) first emerged during the development of instant messaging and presence mechanisms. XCAP is an application-level protocol to allow remote manipulation of data exposed as collections of XML documents. It has been definitively adopted in the NGN Service Layer as the base for the XML Document Management (XDM) technology.

4

Extensible Markup Language Basics

Extensible Markup Language (XML) documents are simple case-sensitive text documents describing structured data. In XML, data are arranged in the form of a tree. Each node of the tree is called an element, which is syntactically represented as a start tag–end tag pair. Tags contain the element name enclosed between the less than (<) and greater than (>) symbols. The end tag also contains a trailing forward slash symbol. Each element may contain text or be empty.

```
<elementName>text</elementName>
<elementName></elementName>
```

Empty elements may also be represented with the following notation:

```
<elementName/>
```

Elements may have attributes in the form of simple key-value pairs. The list of attributes appears inside the start tag next to the element's name. The key is separated from the value by an equals sign, and the value is enclosed in single or double quotation marks:

```
<elementName attributeName1="value1"
attributeName2="value2">text</elementName>
```

Each element may have no more than one attribute with a given name. The names of elements and attributes may contain alphanumerical characters (including non-English letters, numbers, and ideograms), underscores, hyphens, and periods. Other characters are not allowed. Elements may hierarchically contain other elements, may be empty, or may contain text.* Text inside an element is not allowed to

* This mostly occurs in data-oriented documents; however, XML also allows for more free-form documents, with elements containing both text and other elements ("mixed content," similar to Hyptertext Markup Language, HTML).

contain some specially reserved characters such as <, >, &, and the single and double quotation marks that are escaped by, respectively, `<`, `>`, `&`, `'`, and `"`.[*]

XML documents may be commented. Comments are enclosed within two combinations of symbols: `<!--` and `-->`. They may appear anywhere in the document, even before the root element, but not inside a tag or another comment.

```
<!-- This is a comment -->
<elementName attributeName1="value1"
attributeName2="value2">text</elementName>
```

Like comments, processing instructions are enclosed within two combinations of symbols: <? and ?>. They have the following form:

```
<?name instructions?>
```

with name referring to the application or process for which the instructions are intended. However, the name *xml* (in any combination of case) is not permitted, as it is reserved for XML declarations. An XML declaration appears as the first line of an XML document and contains information about the XML version to which the document conforms, about the encoding of the document (by default, UTF-8), and about whether the document needs an external Document Type Definition (DTD) file to determine the correct values for parts of the document that might not be specified within it (e.g., default values for nonspecified attributes). For example:

```
<?xml version="1.0" encoding="ASCII" standalone="yes"?>
```

To avoid mixing elements from different sources that could potentially have the same name, namespaces are introduced. In this way elements' and attributes' names are pairs (namespace, localname). The namespace is in the form of a Uniform Resource Identifier (URI) with a prefix used to avoid repeating the URI throughout the whole XML document. The prefix is bound to the namespace URI through

* Alternatively, it is possible to use a CDATA section, set off by `<![CDATA[` and `]]>`. Everything within a CDATA section is treated as raw character data.

a declaration that appears in the document in the form of an attribute, either on the top element or another element inside the document (in the latter case the namespace has a scope limited to the element and any subelements appearing hierarchically under that element). The pair (namespace, localname) is referred to as the fully qualified name (FQN) and is expressed in the form

```
namespace:localname
```

Namespace and localname can start with a letter or underscore and contain letters, digits, underscores, dashes, and periods. XML documents conforming to the syntactic rules explained above (plus other rules that can be found in Thompson et al. 2004) are said to be "well formed"; that is, they can be parsed by an XML processor.

An XML document is said to be "valid" if it conforms to DTD or XML schema.* DTD and XML schemas are quite common choices but are not the only possibilities.† Both DTD and XML schemas are expressed as text files that contain the grammatical rules—the schema—for the elements and the attributes that appear in a given XML document. Compared with XML schemas, DTDs are less verbose but have less expressiveness.

Any valid XML document is referred to as an *instance* of the schema. A schema is used to define, for example, the names of the elements that might appear in an instance, which attribute what they may or need to have, and which and how elements can be nested. For instance, the following XML document

```
<?xml version = "1.0"?>
<!DOCTYPE note SYSTEM "schema.dtd">
<note>
<text number="2" date="2001-05-22">This is a note.</text>
<text number="3" date="1999-08-18">Another note is here.</text>
</note>
```

is well formed and valid according to the following DTD:

* Note that not all XML documents have an associated DTD or an XML schema.

† There are other schema definition languages. Two examples are the RELAX NG schemas, whose specifications are available online at http://www.oasis-open.org/committees/relax-ng/spec-20011203.html, and the ISO Document Schema Description Languages (DSDLs) specified by ISO/IEC 19757.

```
<!-- filename: schema.dtd -->
<!ELEMENT note (text*)>
<!ELEMENT text (#PCDATA)>
<!ATTLIST text
   number CDATA #REQUIRED
   date CDATA #REQUIRED>
```

With the following slight modification:*

```
<?xml version = "1.0"?>
<note xmlns:xsi="http://www.w3.org/2001/XMLSchema-instance"
  xsi:noNamespaceSchemaLocation="schema.xsd">
<text number="2" date="2001-05-22">This is a note.</text>
<text number="3" date="1999-08-18">Another note is here.</text>
</note>
```

this instance document also validates against the following XML schema:

```
<?xml version="1.0"?>
<!-- filename: schema.xsd -->
<xs:schema xmlns:xs="http://www.w3.org/2001/XMLSchema">
 <xs:element name="note">
  <xs:complexType>
   <xs:sequence>
    <xs:element name="text" minOccurs=1 maxOccurs="unbounded">
     <xs:complexType>
      <xs:simpleContent>
       <xs:extension base="xs:string">
        <xs:attribute name="number" type="xs:positiveInteger"/>
        <xs:attribute name="date" type="xs:date"/>
       </xs:extension>
      </xs:simpleContent>
     </xs:complexType>
    </xs:element>
   </xs:sequence>
  </xs:complexType>
 </xs:element>
</xs:schema>
```

There are some differences to be considered:

* This document contains an attribute xsi:noNamespaceSchemaLocation defined in the namespace http://www.w3.org/2001/XMLSchema-instance. This namespace is used to host attributes that instruct the XML processor to bind the instance document to a schema (or schemas) and namespaces (if any). In the example the XML processor is told that the instance document is associated to an XML schema that can be found in a local file named `schema.xsd`, but it is not associated with any namespace.

- XML schema documents are themselves XML documents and natively support namespaces whereas DTD ones do not.
- XML schema provide the ability to explicitly set the minimum and maximum number of times an element may occur (using the minOccurs and maxOccurs attributes).
- XML schema define a wide type system, containing types that are very similar to those supported by many programming languages. They enable association of these types to elements and attributes.

The XML schema type system offers built-in types that can be referred to from within any schema document using the FQN notation. For example, the *string* type is referred to as `xs:string`, where the prefix `xs`* is associated to the namespace http://www.w3.org/2001/XMLSchema; the *date/time* type is defined as `xs:dateTime` and so on.

A literal (i.e., a text) appearing inside an XML element or attribute associated to a built-in or user-defined type in the corresponding XML schema is called a *typed literal.*

Also, XML schema allows definition of new types, extending the built-in type system. User-defined types are specified directly inside the XML schema using the same syntax used for elements.

Elements and user-defined types in a schema document may be simple or complex. Simple elements may contain only text; complex elements may contain attributes and other elements other than text. XML schema permits the minimum and maximum number of occurrences of nested elements as well as restrictions that limit the range of allowed values for literals to be specified.

An XML schema document is itself a well-formed and valid XML document. The topmost element of this document is xs:schema. To associate a schema to a particular namespace, the attribute xs:targetNamespace is used. This attribute occurs inside the xs:schema element, and its value represents the URI of the namespace. Once a schema defined in an XML schema document is bound to a namespace, that namespace must be used to refer, from within another document, to the elements and the types defined in that schema. For instance,

* Even if the choice of the prefix is free, traditionally FQNs referring to XML schema types are associated with the prefix `xs` (or `xsd`).

if the namespace http://example.org/calendar were associated to the example schema illustrated previously:

```
<xs:schema xmlns:xs="http://www.w3.org/2001/XMLSchema"
targetNamespace="http://example.org/calendar">
...
</xs:schema>
```

the instance document—or even any other schema document—could refer to the element `note` defined in this schema using its FQN. The following example is a modified version of the instance document illustrated and uses the prefix `cal` to refer to the namespace http://example.org/calendar:

```
<?xml version="1.0"?>
<cal:note xmlns:xsi="http://www.w3.org/2001/XMLSchema-instance"
 xsi:schemaLocation="http://example.org/calendar schema.xsd"
 xmlns:cal="http://example.org/calendar">
 <text number="2" date="2001-05-22">This is a note.</text>
 <text number="3" date="1999-08-18">Another note is here.</text>
</cal:note>
```

Also, this example uses the xsi:schemaLocation attribute to specify where the schema can be found. This attribute contains two tokens. The first is a URI specifying the schema "logical" location, that is, the address from which the schema should be retrieved (usually assumed to be the same URI that identifies the schema if the URI is a URL,* in this example http://example.org/calendar). The second specifies its physical location (in this example a file named `schema.xsd`).

By default, all top-level element names appearing in an instance document must be FQNs, whereas nested elements and attribute names are not qualified (thus not prefixed), as in the previous example. It is possible to require explicit qualification for attributes by setting the value of the optional `attributeFormDefault` attribute—appearing in the `xs:schema` element—to "qualified." Similarly for nested elements, the optional `elementFormDefault` attribute is set to "qualified," as in the following example:

* Once more the dualism between identification and location (e.g., Uniform Resource Name [URN] and Uniform Resource Locator [URL]) occurs.

```
<xs:schema xmlns:xs=http://www.w3.org/2001/XMLSchema
 targetNamespace="http://example.org/calendar"
 elementFormDefault="qualified">
 ...
</xs:schema>
```

The corresponding instance document becomes

```
<?xml version="1.0" encoding="UTF-8"?>
<cal:note xmlns:xsi="http://www.w3.org/2001/XMLSchema-instance"
 xsi:schemaLocation="http://example.org/calendar
  schema.xsd"
 xmlns:cal="http://example.org/calendar">
 <cal:text number="2" date="2001-05-22">This is a note.</cal:text>
 <cal:text number="3" date="1999-08-18">Another note is here.
   </cal:text>
</cal:note>
```

In recent years, these features have all contributed to the acceptance of XML schema as the preferred schema definition language, especially for data-oriented documents that are intended to support information exchanges among different software applications and systems over the Internet.

XPath

An addressing mechanism is needed to refer to elements (and attributes) inside an XML document. Even if each element has a name (FQN), there might be more than one element with the same name inside a document, as in the case of elements occurring in a sequence.

XPath (Clark and DeRose 1999) allows addressing any kind of XML node contained in a specified XML document* in a familiar way, using paths that are similar to those used to address files in a hierarchical file system.

An XPath expression starts with a forward slash (which, by itself, identifies the whole document) and is composed of a sequence of FQNs separated by forward slashes. Each FQN represents the name of an element that needs to be traversed to get to the target node. For example, given the XML document

* XPath works on a single input document, which is externally specified.

```
<?xml version="1.0"?>
<note>
<text number="2" date="2001-05-22">This is a note.</text>
<text number="3" date="1999-08-18">Another note is here.</text>
</note>
```

it is possible to address the top-level `<note>` element using the XPath expression

```
/note
```

In general, the evaluation of an XPath expression against an XML document returns a set of nodes. The set may be empty, or it may contain one single node (as in the previous example) or more than one node. When there are multiple occurrences of the same element name, one specific element is selected by including the index of the element that needs to be traversed. Syntactically, the index is given by the element position enclosed in squared brackets and appended to the path. For example, to address the second `text` element inside the `note` element of the previous example, the Xpath

```
/note/text[2]
```

is used. This expression returns

```
<text number="3" date="1999-08-18">Another note is here.</text>
```

Attributes are addressed similarly to elements by prefixing the "at" symbol before the attribute name. For example,

```
/note/text[2]/@number
```

returns "3."

Other reversed symbols in XPath include wildcards (*), which are used to match any element node (regardless of name). Likewise, the combination `@*` matches any attribute. Double forward slash (//) matches all the descendants of a given node at any depth of the tree and is particularly useful when the full path to the node is unknown. For example,

```
//text
```

returns all the `text` elements contained in the document.

As in many file systems, the single dot (.) indicates the current position and the double dot (..) the parent node. For example,

```
//text/..
```

identifies the element `note`, which is the parent of any `text` element found in the document.

The vertical bar (|) indicates alternatives; for example,

```
//(note|text)
```

selects all the `note` elements as well as the `text` elements at any depth.

XPath allows predicates. Each step in a path may contain a predicate that restricts the selection of nodes. Predicates are written as expressions in square brackets. For example,

```
//text[@number>10]
```

selects any `text` element inside a document whose attribute `number` is greater than 10.

Functions and operators may be used inside XPath expressions. There are built-in functions to manipulate strings, numbers, and sets of nodes; to get node properties such as name and position; to access text nodes; and to retrieve comments and processing instructions. XPath 2.0 offers a greater set of functions and operators and uses node sequences (ordered lists of nodes) instead of sets. XPath 2.0 is part of XQuery, the XML query language.

XQuery

XQuery (Boag et al. 2010), a wide extension of XPath, provides a means to define complex SQL-like queries on XML data. XPath is useful to select and extract a subset of nodes from an input XML document. However, it is not possible to manipulate data using XPath, such as changing names or the order of elements in a sequence, adding or removing elements or attributes, and using nodes from external documents. XQuery makes these actions possible.

XQuery deals with XML nodes (documents, elements, attributes, text nodes, comments, and processing instructions) and sequences, that is, ordered lists of possibly duplicated nodes. An XQuery processor is able to, for example, read, process, and return* any sequence of documents, elements, attributes, and text nodes, including collections of full documents and fragments of XML documents.

Variables are allowed. Syntactically, variables are marked with a trailing dollar sign followed by a name, such as $x, and are similar to variables appearing in SQL expressions. XQuery also defines a broad set of built-in functions (e.g., node manipulation, string manipulation, mathematical and logical operators, date and time handlers, and aggregate functions). Custom functions may be defined as well.

A query is made up of two parts: the prologue and the body. The prologue typically contains namespace declarations, variable declarations, or schema import declarations.† Each declaration is terminated by a semicolon. In the following example, a namespace is declared and a variable is bound to an external document. The example also includes comments, enclosed between the symbols `(:` and `:)`.

```
(: Use the following instruction to define the namespace ex :)
declare namespace ex = "http://example.org/";

(: Assign a value to a variable. The doc() function selects a
  document by name :)
declare variable $aDocument := doc("sample-document.xml");
```

The body consists of one or more expressions separated by a comma. Different kinds of expression are allowed including XPath expressions—each valid XPath expression is also a valid query in XQuery—and FLWOR (pronounced "flower") expressions.

FLWOR expressions are the very heart of XQuery. They are named after the five different kinds of clauses they may contain: `for`, `let`, `where`, `order by`, and `return`.

The `for` clause is used to let the XQuery processor iterate through a sequence of nodes. These clauses can be nested. The keyword `let` binds a variable to a value. The value may be a node or an atomic value,

---

* Actually, every expression in XQuery evaluates to a sequence, which may be empty, may contain just one item, or may contain more than one item.

† Despite schema import declarations being optional (the query would work even without them), having them allows the XQuery processor to understand custom types defined in schemas and allows for exceptions when type constraints are not satisfied.

such as a string literal, a numeric literal, or a typed literal.* These values are referred to as *bindings*. Using variables is a convenient way to simplify an expression and make it more readable; variables also boost performances when the query is evaluated. However, the usual trade-off between memory (i.e., number of variables) and execution time has to be considered when designing nontrivial queries. `Where` is an optional clause that allows one to define conditions to filter out nodes. The `return` statement concludes a query and specifies what has to be returned. A query based on an XPath expression normally returns a sequence of nodes; however, using FLWOR expressions, the output is more flexible and the query may also return documents built on the fly. In the latter case, curly brackets are used to distinguish what has to be returned verbatim from what has to be evaluated; expressions inside curly brackets (enclosed expressions) are always evaluated, and the result is concatenated to the rest of text appearing in the return clause. For example, the query

```
(: Assigns a value to a variable. The doc() function selects a
  document by name :)
declare variable $aDocument := doc("sample-document.xml");

(: Following is the actual query :)
for $text in $aDocument/note/text
where $text/@number > 2
return
  <note>
    {$text}
    <author>Alice Smith</author>
  </note>
```

splits the original document into a sequence of nodes and returns only `text` elements whose `number` attribute is greater than 2. Also the query adds an `author` element inside each returned `note` element. A sequence of elements (in this example containing only one element) is returned:

```
<note>
 <text number="3" date="1999-08-18">Another note is here.</text>
 <author>Alice Smith</author>
</note>
```

* Typed literals are actual literals which are recognized as instances of types and handled as such by operators. Typical examples include date/time and Boolean values.

Mixing XML and XQuery constructs is possible. For instance, a whole query may be enclosed in curly brackets and put inside a parent XML element, as in the following example. This example also uses the `order by` clause to reorder the `text` elements in chronological order.

```
declare variable $aDocument:=doc("sample-document.xml");

<doc>
{
for $text in $aDocument/note/text
order by $text/@date
return
 <note>
  {$text}
  <author>Alice Smith</author>
 </note>
}
</doc>
```

The result is a sequence containing one node, the `doc` element, wrapping a sequence of two `note` elements:

```
<doc>
 <note>
  <text number="3" date="1999-08-18">Another note is here.</text>
  <author>Alice Smith</author>
 </note>
 <note>
 <text number="2" date="2001-05-22">This is a note.</text>
  <author>Alice Smith</author>
 </note>
</doc>
```

5
Web Services

Traditionally, web services, or "big web services"—to distinguish them from other services running on the web using different paradigms, such as stateless Representational State Transfer (REST) interfaces—have been popular with the traditional enterprises. A web service is a resource provided by a particular provider described in terms of its allowed operations in a common language, that is, the Web Service Description Language (WSDL). Knowing the WSDL description, a client (consumer) can invoke each operation by sending a request contained in a particularly formatted body. After the execution, a response is returned.

Other than describing the web service, a WSDL description contains information on the particular *transport* protocol used to communicate with the provider (e.g., Hypertext Transfer Protocol [HTTP], Simple Mail Transfer Protocol [SMTP]), the endpoint at which the provider can be contacted, and the message format (typically the Simple Object Access Protocol [SOAP]). Service descriptions may be published in a service registry known as Universal Description, Discovery, and Integration (UDDI) and combined using an executable language such as the Business Process Execution Language (BPEL).

The collection of web service protocols is often represented as a stack on top of the (logical) transport layer (Figure 5.1).

WSDL

WSDL is an Extensible Markup Language (XML)-based language used for describing the interface of a web service (Christensen et al. 2001). A typical WSDL document is composed of several sections.

The type section defines datatypes handled by the service; datatypes are defined using XML schema, which is embedded in the WSDL document under the `wsdl:types` element.

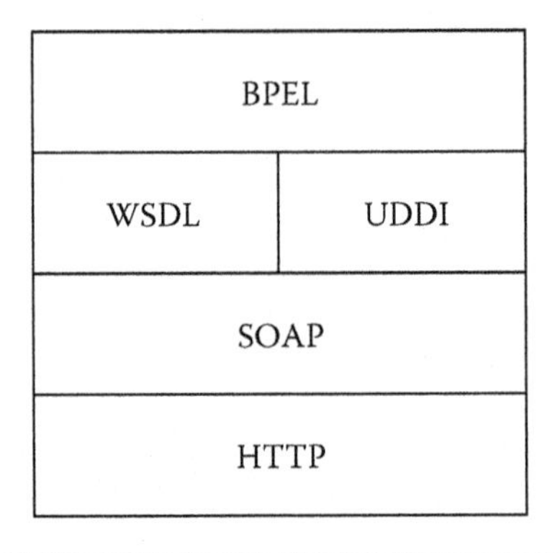

Figure 5.1 The web service stack.

Messages describe abstract messages that are exchanged between the consumer and the provider. Each message contains "message parts" providing the parameters transmitted with the message. The message parameters must match the types defined in the `wsdl:types` section. A message sent from the consumer to the provider is called an *output message*; a message going in the opposite direction is an *input message* or a *fault*.

Messages were removed in WSDL 2.0, which uses XML schema types directly for defining bodies of inputs, outputs, and faults.

Operations describe the actions performed by the services. Operations are executed through WSDL Message exchanges and are arranged into *Port Types* (aka *Interfaces* in WSDL 2.0).

There are four kinds of operations. A *one-way* operation consists of a message sent from the consumer to the provider. In a *request–response* operation, there is one request message from the consumer to the provider followed by a response message in the opposite direction. In a *notification* operation, only one message is sent from the provider to the consumer. A *solicit–response* operation consists of an initial solicitation sent by the provider to the consumer followed by a response issued by the consumer. (Notification and solicit–response are not supported by HTTP-based protocols. Typically, web service developers use polling instead. Another solution, WS-Notification developed by the Organization for the Advancement of Structured Information Standards [OASIS] consortium, will be discussed later.)

Operations that involve bidirectional message exchanges may fail at the provider's side or at the consumer's side. Thus, WSDL provides a third kind of message, *fault*, which can be sent instead of a response to indicate an operation failure. An example of WSDL code, containing the afore described elements, is detailed in the following example:

```
<wsdl:definitions xmlns="http://schemas.xmlsoap.org/wsdl/"
xmlns:wsdl="http://schemas.xmlsoap.org/wsdl/" xmlns:xs="http://www.
w3.org/2001/XMLSchema"
xmlns:soap="http://schemas.xmlsoap.org/wsdl/soap/"
xmlns:soapenc="http://schemas.xmlsoap.org/soap/encoding/"
xmlns:tns="http://example.org/profile"
targetNamespace="http://example.org/profile"
elementFormDefault="qualified"
attributeFormDefault="unqualified">
  <wsdl:types>
    <schema
    targetNamespace="http://example.org/profile"
    xmlns="http://www.w3.org/2001/XMLSchema"
    elementFormDefault="qualified"
    attributeFormDefault="unqualified">
      <simpleType name="ResultCode">
        <restriction base="xs:int">
          <enumeration value=0/>
          <enumeration value=1/>
          <enumeration value="2"/>
        </restriction>
      </simpleType>

      <complexType name="ArrayOfString">
        <complexContent>
          <restriction base="soapenc:Array">
            <attribute ref="soapenc:arrayType" wsdl:arrayType="xs:string[]"/>
          </restriction>
        </complexContent>
      </complexType>
      <complexType name="UserPrefs">
        <sequence>
          <element name="prefID" type="tns:ArrayOfString"/>
          <element name="prefValue" type="tns:ArrayOfString"/>
        </sequence>
      </complexType>
      <complexType name="UserPrefsIDs">
        <sequence>
          <element name="prefID" type="tns:ArrayOfString"/>
        </sequence>
      </complexType>
    </schema>
  </wsdl:types>
  <wsdl:message name="ReturnCodeMessage">
    <wsdl:part name="value" type="tns:ResultCode"/>
  </wsdl:message>
  <wsdl:message name="GetPreferencesMessage">
    <wsdl:part name="username" type="xs:string"/>
    <wsdl:part name="UserPrefsIDs" type="tns:UserPrefsIDs"/>
  </wsdl:message>
  <wsdl:message name="ReturnPreferencesMessage">
```

```
    <wsdl:part name="username" type="xs:string"/>
    <wsdl:part name="UserPrefserences" type="tns:UserPrefs"/>
  </wsdl:message>
  <wsdl:message name="UpdatePreferencesMessage">
    <wsdl:part name="username" type="xs:string"/>
    <wsdl:part name="UserPrefserences" type="tns:UserPrefs"/>
  </wsdl:message>
  <wsdl:portType name="PROFILE">
    <wsdl:operation name="GetPreferencesEvent">
      <wsdl:input message="tns:GetPreferencesMessage"/>
      <wsdl:output message="tns:ReturnPreferencesMessage"/>
      <wsdl:fault name="Value" message="tns:ReturnCodeMessage"/>
    </wsdl:operation>
    <wsdl:operation name="UpdatePreferencesEvent">
      <wsdl:input message="tns:UpdatePreferencesMessage"/>
      <wsdl:output message="tns:ReturnCodeMessage"/>
    </wsdl:operation>
  </wsdl:portType>
</wsdl:definitions>
```

The binding section is used to link the abstract descriptions contained in PortTypes into a concrete format used to transport the message on the wire.

Simple Object Access Protocol

When binding a web service to a concrete transport protocol, SOAP (Gudgin et al. 2007) is a common choice. SOAP is XML based and defines a structure made of an envelope containing one (or more) optional header and a body. The envelope contains information such as which methods to invoke, optional parameters, return values, and faults, if any. Each header may contain additional information that may be used by intermediary nodes handling the SOAP message between the consumer and the provider. The body contains the actual data that need to be exchanged. An example of SOAP binding is detailed as follows (with part of the document omitted):

```
<wsdl:definitions … >
  …
  <wsdl:binding name="ProfileBinding" type="tns:PROFILE">
    <soap:binding style="rpc" transport="http://schemas.xmlsoap.org/
      soap/http"/>
    <wsdl:operation name="GetPreferencesEvent">
      <soap:operation soapAction="http://example.org/profile/
        GetPreferencesEvent"/>
      <input>
        <soap:body use="encoded" namespace="http://example.org/profile"/>
      </input>
      <output>
```

```
      <soap:body use="encoded" namespace="http://example.org/profile"/>
    </output>
    <fault name="Value">
      <soap:body use="encoded" namespace="http://example.org/profile"/>
    </fault>
  </wsdl:operation>
  <wsdl:operation name="UpdatePreferencesEvent">
    <soap:operation soapAction="http://example.org/profile/
      UpdatePreferencesEvent"/>
    <input>
      <soap:body use="encoded" namespace="http://example.org/profile"/>
    </input>
    <output>
      <soap:body use="encoded" namespace="http://example.org/profile"/>
    </output>
  </wsdl:operation>
 </wsdl:binding>
</wsdl:definitions>
```

The SOAP binding element specifies the transport to be used (HTTP, as in the example is common but in theory alternative transports such as SMTP are viable as well) and the style. SOAP provides two choices: RPC and document (Table 5.1). With RPC, the SOAP body will wrap the parameters to be passed inside an element containing the name of the web method called for. When *document* is used, the SOAP body will contain only the parameters to be passed.

The SOAP body binding element provides information on how to assemble the different parts of the abstract WSDL message inside the body element of the SOAP message. The parts are specified inside the *part* attribute, which is optional. If not present, by default, the SOAP body will contain all parts of the WSDL message. The required *use* attribute indicates whether the message parts are encoded (Table 5.2). When *encoded* is used, each message part is mapped into an element of the body with a *type* attribute specifying its datatype. When *literal* is used, the parts appear as an XML document instance conforming to the XML schema defined in the type section of the WSDL document.

Table 5.1 Possible Values for the SOAP Style Attribute

STYLE	
RPC	The SOAP body contains an element with the name of the web method being invoked (which, in turn, contains an entry for each parameter and the return value of this method).
document	The information containing the method's name is not transported in the SOAP body (but it is typically obtained from the URL), only the message parts are.

Table 5.2 Possible Values for the SOAP Use Attribute

USE	
encoded	Each message part is mapped into an element with a `type` attribute specifying its datatype. Example: `<Domain xsi:type="xs:string">example.org</Domain>`
literal	Data appear as an XML document instance conforming to the XML schema defined in the Type section of the WSDL document. Example: `<Domain>example.org</Domain>`

Even if four combinations are possible in theory, only two of them are prevalent: RPC/encoding and document/literal. When using RPC/encoding, there is a clear overload due to the specification of the type of each message part, but this is the only solution in the case of polymorphic methods, that is, methods whose parameters may be of different types (e.g., a field *date* may be of type xs:string or xs:date). With document/literal the method name is not supplied in the body and should be obtained otherwise.

In WSDL, a given WSDL binding coupled with an endpoint forms a port (*Endpoint* in WSDL 2.0). A web service is then a collection of ports. The following example illustrates the definition of a possible web service (with some definitions omitted):

```
<wsdl:definitions … >
  …
   <wsdl:service name="ProfileService">
     <documentation>The User Profile Web Service allows you to retrieve
       and modify your user preferences.</documentation>
     <port name="ProfilePort" binding="tns:ProfileBinding">
       <soap:address location="http://example.org/profile/endpoint"/>
     </port>
   </wsdl:service>
</wsdl:definitions>
```

Web Service Repository and Orchestration

Service descriptions may be published in a service registry known as UDDI (Clement et al. 2004). UDDI is typically encountered as a web service or in the form of a human-browsable Hypertext Markup Language (HTML) page. The UDDI is composed of three sections: (1) white pages, containing general information about the service; (2) yellow pages, classifying services according to the type of business they perform; and (3) green pages, providing technical information about the service (i.e., the service provider's contacts for technical assistance). UDDI systems were originally modeled as open brokers available on the Internet scale where consumers could obtain services and combine them using an executable language such as BPEL (Alves et al. 2007).

Public UDDI nodes supported by well-known information technology (IT) organizations were in place until 2006 but then were discontinued (SAP News Desk 2005). Today this paradigm appears less fascinating to developers who mainly shifted toward REST architectures. UDDI systems, however, remain in place in most organizations and are mainly used to support orchestration of local services.

Notification and Addressing

WSDL describes notifications as one of four core types of operations, but many implementations do not support them.* To overcome this issue, OASIS released a set of specifications known as WS-Notification (Graham, Hull, and Murray 2006). WS-Notification describes a mechanism enabling a notification consumer to subscribe to a subset of notification events and receive them as asynchronous messages.

WS-Notification leverage on WS-Addressing is a standard developed by W3C (Gudgin, Hadley, and Rogers, 2006). Using WS-Addressing, a single SOAP interaction may be spread across multiple HTTP interactions. This is achieved by enriching the SOAP header with a set of message addressing properties containing the endpoint references (EPRs) to which the reply message or the fault message must be returned.

Addressing, particularly client addressing, proved to be a critical problem behind the introduction of notification capability. A later study focusing on the real needs of a large telecom company, as part of the work performed on the OASIS Service Oriented Architecture for Telecom Technical Committee (SOATEL TC), recently highlighted further requirements. One key finding of SOATEL TC was that addressing should be revised to explicitly allow intermediaries between the notification initiator and the recipient.

* For example, in AXIS—one of the most widely used tools for handling web services in the Java 2 Enterprise Edition environment—notifications are explicitly disabled. Attempting to generate Java code from a WSDL document containing notifications results in a warning stating that notification operations are not supported by the implementation.

6

XML Configuration Access Protocol

Jonathan Rosenberg first proposed the Extensible Markup Language (XML) Configuration Access Protocol (XCAP) in 2003. At that time, Rosenberg was working to develop instant messaging and presence mechanisms. As part of his work, he developed an application-level protocol to allow remote manipulation of data exposed as collections of XML documents. In the following years, XCAP became the core of the XML Document Management (XDM) technology, used in next-generation networks (NGNs) to manage several user and group data. XDM added subscription and notification capabilities to XCAP, using traditional Session Initiation Protocol (SIP) messages, and many more features; however, the syntax and the convention used were the same ones originally developed for XCAP.

XCAP operates on information resources represented as XML statements that contain "per-user" information. Examples include, but are not limited to, user profiles, buddy lists in instant messaging applications, and preferences related to presence information such as "show my state as busy."* To refer to a set of XML statements, it is common to use the expression XML document; however, this does not necessarily imply that they are physically stored as a file in a file system, as the term *document* might suggest. Native XML databases, as opposed to traditional relational databases, are able to store and operate on collections of XML documents as well. XCAP simply assumes that XML documents are information resources that reside in a network server; however, they can be managed directly by the user through a client. In principle, Hypertext Transfer Protocol (HTTP) methods alone could be used to directly manipulate these resources, simply

* Actually, XCAP has emerged mainly as a solution to typical problems that the first instant messaging and presence applications were facing; therefore, it is common to refer to use cases from that application domain.

assuming that a Uniform Resource Indicator (URI) is associated with the *target* resource. This approach is similar to the first usage that early web developers made of HTTP, where a single change in an element of a Hypertext Markup Language (HTML) page was rendered by reloading the whole page. Obviously this approach had the drawback of the overhead involved in fetching a whole page when it would really be necessary to transmit only differential changes. The need to make differential changes is even more critical when pages are replaced with resources such as XML documents, which contain dynamic data that are frequently modified, and if the protocol is used on a wireless interface. Therefore, it is essential to allow the manipulation of partial pieces of the data, that is, single XML tags and attributes. XCAP solves this problem by introducing a set of conventions to map XML documents and components thereof into HTTP URIs. It also introduces data validation constraints, rules describing how mutually dependent resources should be modified and access authorization policies.

XCAP Addressing

XCAP combines the addressing power of XPath together with the semantics of HTTP methods to provide full XML document management capabilities. As described, HTTP methods already allow full management in terms of create, read, update, and delete operations of resources identified by URIs. What is interesting to be noted is that a "resource" does not need to be necessarily a file; it could be a finer granule entity, such as an element or attribute in an XML document deployed somewhere on the network. To this end the semantics of the request URI in the HTTP request message is opaque to the HTTP protocol itself (Jacobs and Walsh 2004) but must be understood at the endpoint that accepts the HTTP request. In particular, an XCAP endpoint accepts HTTP requests containing request URIs structured as follows: the next segment in the document selector path is used to distinguish between global resources and user-specific resources. The URI begins with the XCAP root, followed by a document selector and a node selector. Typically the XCAP root is a simple URI identifying the host running the XCAP server. The document selector is a sequence of path segments, separated by a slash ("/"). Since the first XCAP server may manage several different unique usages, the first path segment, called Application Unique Identifier (AUID), defines which application usage the request refers to. Obviously, user-specific resources contain users' settings, whereas global resources are for global settings. This segment consists of either the token "global" or the token "users." In the first case, the segment is followed directly by the identifier of the specific document; in the latter case the segment is followed by an additional segment that identifies the specific user. Each user is in fact associated with a specific identifier called XCAP user identifier (XUI). In SIP applications the XUI is generally the address of record (AoR) associated with the SIP user. Characters that are not valid in a HTTP URI but are valid for an AoR (question marks and slashes) are percent encoded. Finally, the document identifier is provided. The document identifier can be either a single token or a path-like structure, which does not necessarily represent the physical directory structure on the server where the document resides. If a directory structure is chosen, there is an additional constraint specifying that there must be no documents with the same name as a directory name. However, the directory structure for the document identifier is discouraged not to disclose the server's internal file structures.

The following part of the XCAP URI, the node selector, is separated from the document selector by a double tilde (~~) and consists of an expression that identifies an element, an attribute, or a set of namespace bindings inside the selected document. The node selector can be empty; in this case the URI identifies the whole XML document. The node selector can also consist of an expression syntactically conforming to XPath, except that disallowed characters will be percent encoded (in particular, squared brackets and quotes, which are not permitted in a URI, are percent encoded; also, if the XML document uses a non–UTF-8 encoding for its elements and attributes, then these need to be converted in UTF-8 octects and percent encoded). Figure 6.1 illustrates the structure of an XCAP URI. For the sake of simplicity, the URI has not been percent encoded.

The corresponding percent encoded expression is

```
http://xcap.example.com/serv-name/users/sip:user@example.com/index/~~
/serv-name/entry%5b@id=%221234567%22%5d
```

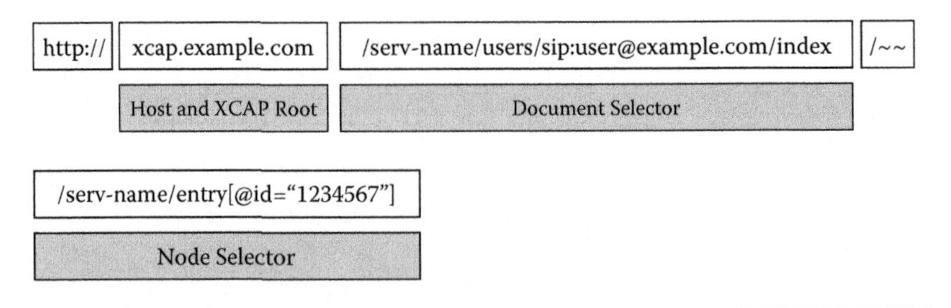

Figure 6.1 An XCAP URI broken down into its parts (wrapped into two lines for readability).

Not all XPath expressions are supported by XCAP; the supported ones allow an element (or attribute) to be selected by specifying, in each step of the path, nodes matching one or more of these three conditions: name, position, and attribute value. Wildcards are supported in making the selections. Selection by name or attribute value is usually the preferred choice, because the value of the attribute or the name of the element is usually more meaningful to the application client than the position of the node inside the target document; however, selection by position combined with selection by attribute value is also a frequent event. In PUT requests this technique is called *positional insertion*. An example of positional selector is the following:

```
http://xcap.example.com/serv-name/users/sip:user@example.com/index/~~
/serv-name/container-entry/*[2][@name="foo"]
```

This selector picks the second subelement within the *container-entry* element whose attribute name has value "foo."

Comments, text content, whitespace, and processing instruction, either originally present in XML documents or added later, are preserved during XCAP operations; nevertheless, XCAP treats them as annotations and does not define any addressing mechanism for them. XCAP clients are instead limited to manage only XML elements and attributes.

It is noteworthy that to find the correct element or attribute, the XML document is traversed using the node selector expression defined in the XCAP URI. Names, whenever present, are always referred to by the fully qualified names (FQNs) of the elements or attributes present in the document; therefore, the node selector expression also needs to convey information about the namespace bindings that it uses to address the specific node. To report namespace prefixes, XCAP uses an XPointer (DeRose et al. 2011) expression from the `xmlns()` scheme. This expression must appear in the query component of the URI. The following example illustrates the usage of namespaces inside an XCAP URI. Assuming that the following target document is located at http://xcap.example.com/serv-name/users/sip:user@example.com/index,

```
<?xml version="1.0" encoding="UTF-8"?>
<container>
  <ns1:entry xmlns:ns1="urn:example:ns1">
    <element/>
    <ns2:element xmlns:ns2="urn:example:ns2"/>
  </ns1:entry>
</container>
```

the XCAP expression

```
http://xcap.example.com/serv-name/users/sip:user@example.com/index/~~
/container/p:entry/element?xmlns(p=urn:example:ns1)
```

selects the first `element` node appearing under the `p:entry` node. Note that the topmost node, `container`, does not need to be qualified, as it is part of the default namespace in the target XML document. The following URI can be used to select the second `element` node, which is under a different namespace.

```
http://xcap.example.com/serv-name/users/sip:user@example.com/index/~~
/container/p:entry/q:element?xmlns(p=urn:example:ns1)xmlns
(q=urn:example:ns2)
```

It should be noted that FQN prefixes may be different in the request URI and in the returned response. Whereas namespace bindings in the query component of the request URI are defined by the client, those returned by the server are the actual prefixes used in the original document.

If a namespace is not recognized, then the server rejects the request and responds with an error (HTTP code 400).

XCAP and HTTP Methods

An XCAP request is just a simple HTTP request with a request URI that conforms to the previously explained structure. An XCAP server is therefore a compliant HTTP origin server hosting the XCAP application. The server must be aware of the application usage; that is, it must know the schema of the XML documents it stores,* the uniqueness constraints, and all other rules defined by the application usage. Upon receiving the XCAP request, the XCAP server inspects the request URI and uses the document selector to identify the target document.

If the URI refers to an unknown AUID or XUI, then the server must reject the request using the 404 Not Found error code.

The server then looks at the node selector part. If the node selector is empty, then the HTTP request method will apply to the whole document; otherwise, it will apply to the node pointed to by the node selector. The server then inspects the HTTP method invoked to find out which operation has to be carried out. Allowed HTTP methods include GET, DELETE, and PUT. Their semantics are exactly the standard semantics defined in Representational State Transfer (REST) architectures: retrieving (GET), deleting (DELETE), and adding or modifying (PUT). The method POST and other HTTP methods are not used by XCAP specifications.

Before invoking an operation, the client should optionally check that successful results do not violate any constraint defined by the application usage;† in particular, it should check that the XML schema, to which the target document conforms, is maintained and that any uniqueness constraints for values are respected. This check is recommended. Even if it is the server's responsibility to ensure that the application usage constraints are maintained, any precheck performed by the client could help to save bandwidth (by avoiding making requests that will surely fail). This may be critical when the requests are transmitted on a wireless connection or any other bandwidth constrained bearer.

Idempotence is an important property that XCAP exploits by introducing further constraints. HTTP methods GET, PUT, and

---

* There is a minor exception to this behavior that is considered whenever the schemas of the stored XML documents allow for elements from other schemas not defined in the application usage. In this case, it is obvious that the XCAP server is not responsible for checking the schema compliance for the part of the document that allows these a priori unknown elements.

† This is generally possible because XCAP specifications allow and recommend that clients cache the documents (whole or part thereof) in which they are interested. Change of entity tags (as described later) will indicate to the client that the local cached copy is out of sync and that the client needs to obtain a new copy of the document from the XCAP server.

DELETE own this property: when the same PUT or DELETE request is performed twice, the side effects generated by the second request are the same as the side effects of the first one. However, in addition to idempotence, XCAP requires one more constraint; Rosenberg described this principle in short using the compact formula:

```
GET(PUT(x)) = = x
```

This might seem trivial; however, in some events this constraint might be not respected. A typical example is when the content of the request does not match the last part of the node selector. This is illustrated by the following request, which will fail because the Request URI contains a final path segment with an attribute that does not match the value of the correspondent attribute in the element contained in the body of the request.

```
PUT
/serv-name/users/sip:user@example.com/index/~~/serv-name/entry%5b@id=
%221234567%22%5d
HTTP/1.1
Content-Type:application/xcap-el+xml
Host: xcap.example.com
<entry id="7654321">
   <empty/>
</entry>
```

In this case, the server would reject the request indicating a "cannot insert" error. Note that the example above could correspond to two different scenarios. When the original XML document does not present any element that matches the URI contained in the request, the example clearly shows a mistake; however, if the document contains such an element, then the request could be intended as a replacement with a new element keeping a different attribute. However, this request cannot be performed, due to the GET(PUT(x)) = = x constraint. This intended effect can be achieved by a number of workarounds. For example, positional selectors can be used; or it is possible for the client to first retrieve the parent node of the node containing the attribute, examine its content and then locally replace the target element. Finally, the client performs a PUT request containing the new parent node together with all children nodes, including the replaced one and all its siblings.

Similarly, to maintain idempotence, the DELETE method also has some constraints that must be respected. In particular, once an element has been deleted, it cannot be retrieved. The symbolic expression of this constraint is the equation GET(DELETE(x)) = = NULL. The implication is that a positional deletion (i.e., a delete operation invoked on a node selected by position) is possible only whenever the position fits the last element in a sequence; it is easy to see that if this would not happen then the constraint expressed in the previous equation would no longer be valid, as the next element would take the position left empty by the deleted one (unless, as previously stated, the element is the last element in the sequence).

As previously mentioned, the PUT method has a dual usage, as it is used for both modifying and adding new nodes to the target XML document. The action is implicitly selected by the server upon inspecting the request URI. If the request URI points at an existing resource, then its content is modified, creating a new variant of the resource. Otherwise, the resource is created. It is worth noting that the URI present in the request should exactly match the document, the element, or the attribute to be added or replaced.

To perform an insertion or a replacement, upon receiving the PUT request, after having checked the authorization permissions, the server removes the last segment from the request URI, obtaining the so-called *parent URI.* If the URI does not have a node selector, and consequently does not have a node selector separator, then the operation is intended to affect one entire document. In this case the parent URI refers either to the global directory or to the user's home directory (or one of their subdirectories, if the application uses subdirectories). If the parent URI refers to a nonexistent directory, a 409 response is sent back to the client, indicating that the parent of the target resource does not exist. Similarly, if the URI contains a node selector and the parent URI refers to an element inside an XML document that does not exist, a 409 response is also returned. If the parent URI points to a valid XCAP resource, then the server proceeds with validating the content inside the PUT request. In particular, the server checks whether the encoding is UTF-8 (the only encoding accepted by XCAP servers) and whether the Content-Type header field of the request matches the MIME type specified by the application usage for a document or is equal to `application/xcap-el+xml` for an element or `application/xcap-att+xml` for an attribute. If the content consists of an XML element, the server also verifies that it is a well-formed XML fragment. Detailed error response codes are reported in RFC 4825 (Rosenberg 2007a).

The insertion of an XML element is constrained by a precise logic that keeps elements with the same FQN closely together, respecting XML schemas that define a set of sequences one after another. In particular, if there are no other sibling elements with the same FQN and the insertion is not a positional insertion, the element is inserted as the last element among all other sibling elements. If there are siblings with the same FQN, the insertion occurs just after all other siblings with the same FQN but before siblings with any other FQN. When a positional insertion is being made, say at position n, the element is inserted such that there are exactly $n - 1$ elements (with the same FQN, if a FQN has been specified, or element with any FQN if wildcard is used) before it. Namespace declarations contained in an element insertion request content are retained even if they result in redundant namespace declarations for the document.

After the element has been tentatively inserted or modified, the server performs a validation of the whole document, checking whether all constraints defined by the application usage (in particular, schema constraints and uniqueness constraints) are still respected. Violations of one or more of these constraint will result in a failure conveyed to the client with a 409 response containing details on the violated constraints.

Deletion is performed similarly, removing the document, element, or attribute pointed to by the request URI. It is not possible to delete a namespace binding set; therefore, requests of this kind are rejected with a 405 (Method Not Allowed) response code.

The deletion of an element always preserves any surrounding entities, including text, whitespace, processing instructions, and comment.

To preserve idempotence, the server rejects any deletion request that causes another element in the document to be pointed to by the request URI after the deletion has been performed. (As anticipated, this is the reason that positional deletion is permitted only on the last element of a sequence.)

Retrieval of an XCAP resource is, finally, the simplest case. If the request URI has no node selector (and consequently does not have a node selector separator), then the client intends to retrieve a complete XML document that is returned with a 200 OK response if it exists or a 404 response is returned if the resource does not exist. The content type of the body is the MIME type defined by the application usage. If the request URI has a node selector part, then, assuming the operation is successful, the corresponding element, attribute, or namespace binding set is returned in a 200 OK response. MIME types for these elements are, respectively, `application/xcap-el+xml`, `application/xcap-att+xml`, and `application/xcap-ns+xml`. If the element, attribute, or namespace binding is not found, then a 404 response is returned.

Conditional Operations

Conditional operations are possible in XCAP by exploiting entity tags and the If-Match and If-None-Match header fields defined for the HTTP request header. An XCAP server must maintain entity tags for all resources that it maintains. There is a single entity tag for each XML document the server maintains. Elements and attributes inside the document, which are also (XCAP) resources, do share the same entity tag as the document to which they belong.

When an XCAP client starts up, it usually fetches the whole document from the XCAP server and caches it in the local storage. Together with the document itself, the server transmits the entity tag of the document to the client in the HTTP response. In general, an entity tag is always present in the server's 200 or 201 responses following any GET, PUT, or DELETE operation (except when the whole document is deleted).

Using this entity tag and the conditional header, the client may ask the server to perform a specific operation (insertion, deletion, or modification) if and only if it has checked that the local cached copy has not been modified by other clients. If the condition expressed in the conditional header is not matched, the server will respond with an error (412). The client is then advised to discard the old cached document and retrieve a new fresh copy of it (together with its new entity tag). Alternatively, if the conditions are met, the server returns a success response, which also contains the new entity tag of the document (which may be recomputed according to the requested modification). The client should then update the cached document and store the new entity tag for subsequent conditional requests.

Error Handling

XCAP servers may return errors as responses to a client request. The returned errors, classified using standard HTTP response error codes, are XCAP specific. XCAP errors are returned in a HTTP response in which the Content-Type entity-header field is set to the MIME type format `application/xcap-error+xml`, a MIME type explicitly defined for XCAP errors. The entity body returned is an XML document containing a description of the error. Most frequently occurring errors include 404 Not Found, 409 Conflict, and 412 Precondition Failed. The 404 Not Found error indicates that the document, element, or attribute to which the URI refers does not exist. Error 412 is returned on conditional requests using the If-Match or If-None-Match header field defined for the HTTP request header. The 409 Conflict is generated when the operation results in a document that is not well formed or valid, when the resource identified in a request corresponds to multiple elements or attributes, or when then client provided data that violate a uniqueness requirement. This error is also returned whenever the client attempts to set for the second time a value that already exists (e.g., in a list) and is supposed to be unique.

Authentication between the client and the server uses HTTP Digest (RFC 2617, Franks et al. 1999) and is accomplished using HTTP over TLS (Rescorla 2000). The XCAP root URI should therefore be a Hypertext Transfer Protocol Secure (HTTPS) URI. Authorization policies are in general defined by the specific application usage; however, XCAP defines a default authorization with the following policy rules:

- Enable each user to read, add, delete, and modify all XML documents and parts thereof below their home directory
- Enable each user to read XML documents within the global directory
- Enable only trusted users, explicitly provisioned into the server, to read, add, delete, and modify XML documents within the global directory

Exposing Application Usage Capabilities: XCAP-CAPS

XCAP can be extended through the addition of new application usages. Even the core protocol could be extended by defining so-called XCAP extensions. To make the most efficient use of the protocol, XCAP clients need to be aware not only of the semantics of the application they manage but also of the server capability. For example, an application may conform to many XML schema; however, XCAP servers might be required to understand only some of them. Also, there may be different XCAP servers, supporting different XCAP extensions.

To this purpose, a specific XCAP application usage whose AUID is xcap-caps (XCAP server capabilities) has been defined. All XCAP servers need to support at least this application usage, which defines a single XML document available in the global directory listing the AUIDs, extensions, and namespaces understood by the XCAP server. This XML document has a schema defined in the target namespace `urn:ietf:params:xml:ns:xcap-caps`.

7
Open Data Protocol

The Open Data Protocol (OData) is an ongoing effort* to define a universal application-level protocol to access any kind of data available on the web. According to its designers, OData is considered *open* not only because it is designed through an open process and open license but also because it opens silos of data to the web (Pizzo 2012).

OData is based on the Representational State Transfer (REST) architecture and defines conventions, rules, and formats for handling data on the web using Hypertext Transfer Protocol (HTTP) requests. Querying and editing data are both supported by the protocol.

Entity Data Model

OData defines an abstract data model called Entity Data Model (EDM). Each OData service must explicitly expose an Extensible Markup Language (XML) document conforming to the EDM. This document is called the service's metadata document and contains the description of the data in terms of the elements defined by the EDM (namely, entities, properties, relationships, types, and operations). By default, the metadata document has the following location:

```
http://<odata-service-root-uri>/$metadata
```

where <odata-service-root-uri> represents the Uniform Resource Identifier (URI) assigned to the OData service. The client can request the metadata document by issuing a HTTP GET request to that URI

* The text describes the protocol as of its third version, OData V3, from the OData community. Starting from this version at the time of writing, the OASIS OData Technical Committee is in charge of developing the new OData specifications that will become an OASIS Standard. The namespaces used in the examples are the new ones suggested by the Technical Committee members.

Table 7.1 OData System Query Options

SYSTEM QUERY OPTION	MEANING
`$filter=<filter-expression>`	Only entities that match the filter expression are returned. The filter expression supports logical and arithmetic operators, functions, and grouping.
`$expand=<nav-prop-list>`	Normally, for each returned entity, navigation properties are "deferred"; that is, the related entities are returned as URI references. Navigation properties specified as comma-separated values in the `<nav-prop-list>` are instead expanded; that is, related entities are described in line within the returned response.
`$select=<field-list>`	For each entity, return only fields (i.e., properties and operations) explicitly appearing in the comma-separated list of fields `<field-list>`. This option may be combined with `$expand` to limit the number of in-line related entities.
`$orderby=<prop-list>`	The order of returned entities is determined by the comma-separated properties appearing in the `<prop-list>`. For each of them it is possible to specify the order as ascending or descending.
`$top=<n>`	Return only the first n entities.
`$skip=<n>`	Skip the first n entities.
`$inlinecount=allpages|none`	If `allpages` is specified, return the total number of entities matching the request (along with the query result); if `none` is specified, no count is returned.
`$format=<mime-type>`	Same effect as HTTP Accept Request Header field.

and can use the Accept request header to specify the desired format that is by default, XML.*

Alternatively, a second way to define format is by adding a query component to the URI specifying the `$format` System Query Option (Table 7.1). This takes precedence over the Accept request header and is useful when HTTP request header fields are not transmitted with the outgoing request (e.g., due to local firewall limitations).

* XML-based metadata documents conform to an XML schema called Common Schema Definition Language (CSDL). The CSDL is one possible XML-based representation of the abstract EDM.

By parsing the metadata document a client can determine how to interact with the service.

Like a record in a relational database, an OData entity is a structured type uniquely identified by a key. An entity may have properties and relationships (aka *navigation properties*) with other entities. An entity's key consists of a subset of properties whose values collectively identify the entity uniquely. Furthermore, as with tables collecting records in a database, entities are grouped into entity sets. An OData service is then seen as a *container* of entity sets and operations (i.e., the custom business logic) allowed on them, on their entities, properties, and relationships.*

Interacting with OData Services

To provide the client with a starting point, each OData service provides a list of all top-level exposed entity sets called the *service document.* Conventionally, each client can request the service document by issuing a HTTP GET request to the service's root URI, such as

```
GET http://example.org/odata/v3/ HTTP/1.1
```

The format of the response can be specified by using the Accept request header or by appending the `$format` option to the URI. Currently OData defines two formats: OData Atom (application/atom+xml) and OData JSON (JavaScript Object Notation; application/json). The default format is OData Atom.

The Atom format (RFC 4287, Nottingham and Sayre 2005) was originally created to support syndication of web content. An Atom document is an XML-based document composed of elements called *feeds.* Each feed may contain a number of extensible *entries.* OData Atom format leverages on the Atom format and on the Atom Publishing Protocol (RFC 5023, Gregorio and de hOra 2007), which defines how

* OData specifications distinguish operations into actions and functions. Actions may have side effects and may return data. Functions must return data and do not have side effects. Functions, contrary to operations, may be composed. Actions and functions may operate on single entities or on entity sets or they may have a global scope; in the last case they are called *service operations.*

Atom-formatted representations within HTTP request and response messages are used to manipulate resources on the web.

JSON is a much simpler data format than Atom. Natively implemented in the JavaScript programming language, JSON is now supported by almost all modern programming languages, because of its easy syntax and compactness (compared with XML representations). JSON supports only two structures: a set of unordered name–value pairs, called *JSON object*; and an ordered list of values, known as *JSON array*. Contrary to XML, JSON does not support namespaces and JSON documents are schemaless.*

OData Entities

Most of OData specifications regard addressing rules that define how entity sets, single entities, relationships, and properties are identified. An entity is identified by its key within the entity set to which it belongs. For example, assuming that there exists an entity set named *Papers*, and within this set the entity key is represented by a property `Id`, whose type is integer, then a client can retrieve the entity contained in that set by simply issuing a HTTP request:

```
GET http://example.org/odata/v3/Papers(Id=5) HTTP/1.1
```

The first part of the URI (http://services.odata.org/OData) identifies the service root URI; the following is called the *resource path.*

If the resource key is composed of a single property, the property name may be omitted:

```
GET http://example.org/odata/Papers(5) HTTP/1.1
```

but if the key is composed of more than one property, then all of the properties values must be explicitly expressed:

```
GET http://example.org/odata/Papers(ID=5,Year=2008) HTTP/1.1
```

This way of addressing an entity is referred to as *canonical* because other URIs may be assigned to the same entity. For example, an

* Efforts toward the definition of a standard notation for JSON schemas are ongoing (see, for example, Court 2010).

entity may be obtained by invoking a function, an action, or a service operation.

The OData service returns a response according to the semantics of the HTTP protocol. Assuming the resource exists, the response can be a 200 OK message containing the resource serialized according to the format the client has specified (either by the Accept request-header field or the `$format` System Query Option). In OData Atom format, each entity is mapped into an entry.

```
&lt;?xml version="1.0" encoding="utf-8" ?&gt;
<entry xml:base="http://example.org/odata/v3/"
xmlns:data="http://docs.oasis-open.org/odata/ns/dataservices"
xmlns:metadata="http://docs.oasis-open.org/odata/ns/dataservices/
metadata"
xmlns="http://www.w3.org/2005/Atom">
  &lt;id&gt;http://example.org/odata/v3/Papers(5)&lt;/id&gt;
  &lt;title type="text"&gt;An Example Paper&lt;/title&gt;
  &lt;summary type="html"&gt;
This paper contains examples of OData query requests.
  &lt;/summary&gt;
  &lt;updated&gt;2012-01-02T11:12:00Z&lt;/updated&gt;
  &lt;author&gt;
    &lt;name /&gt;
  &lt;/author&gt;
  &lt;link rel="edit" title="Paper" href="Papers(5)" /&gt;
  <link rel="http://docs.oasis-open.org/odata/ns/related/Review"
type="application/atom+xml;type=entry" title="Review"
href="Papers(5)/Review"/>
  <category term="Example.Paper" scheme="http://docs.oasis-open.
org/odata/ns/scheme"/>
  &lt;content type="application/xml"&gt;
    &lt;metadata:properties&gt;
      &lt;data:Id m:type="Edm.Int32"&gt;5&lt;/data:Id&gt;
      &lt;data:Year metadata:type="Edm.Int16"&gt;2008&lt;/data:Year&gt;
      &lt;data:Name metadata:type="Edm.String"&gt;An Example Paper</
data:Name>
      &lt;data:Author metadata:type="Example.AuthorType"&gt;
           &lt;data:AuthorName metadata:type="Edm.String"&gt;Alice
Smith&lt;/data:AuthorName&gt;
           <data:AuthorAffiliation metadata:type="Edm.String"
metadata:null="true"/>
      &lt;/data:Author&gt;
    &lt;data:BestPaperAwarded m:type="Edm.Boolean"&gt;false</
data:BestPaperAwarded>
    &lt;/metadata:properties&gt;
  &lt;/content&gt;
&lt;/entry&gt;
```

The id element in the Atom namespace (http://www.w3.org/2005/Atom) defines the globally unique identifier for the entry. The Atom link elements are used for two purposes. The first use is defining a URI where the entity can be retrieved (a self link with the rel attribute set to "self") or modified (an edit link with the rel attribute set to "edit"). The URI is specified in the href attribute and is relative to the base URI which is set in the xml:base attribute. (In the previous example xml:base appears as an attribute of the entry element.)

The second use of the link element is to state relationships. In the example, the rel attribute specifies the identifier of the relationship,* and the href attribute, depending on the cardinality of the relationship, contains the URI of a single related entity or a collection of related entities. Since an entity set is mapped into a feed, the type attribute will be set to application/atom+xml;type = entry or application/atom+xml;type = feed, respectively. When navigation properties are expanded—as per a request containing the $expand System Query Option (Table 7.1)—related entities are not referenced through the href attribute but appear inline inside a child metadata:inline element containing an Atom entry or an Atom feed, depending on the cardinality of the relationship.

The category element is used to specify the entity type. The content element hosts the OData metadata:property element containing the entity's properties. Each element name within the OData Data Namespace matches the name of a property. For simple typed properties, data values appear directly inside the element, and complex typed properties are encoded as nested elements.

A JSON representation for the same entity may be obtained using the Accept request header field or, alternatively, by appending the $format = json System Query Option to the request URI, that is, the request

```
GET http://example.org/odata/v3/Papers(Id=5)?$format=json HTTP/1.1
```

which returns

* The current OData specifications mandate that "this identifier is a URI made up of the name of the OData Data Namespace, followed by the string /related/ and the name of the navigation property on the entity."

```
{
   "d":{
      "__metadata":{
         "uri":"http://example.org/odata/v3/Papers(5)",
         "type":"Example.Paper",
         "edit_media":" http://example.org/odata/v3/Papers(5)"
      },
      "Review":{
         "__deferred":{
            "uri":" http://example.org/odata/v3/Papers(5)/Review"
         }
      },
      "Id":"5",
      "Year":"2008",
      "Name":"An Example Paper",
      "Author":{
         "__metadata":{
            "type":"Example.AuthorType"
         },
         "AuthorName":"Alice Smith",
         "AuthorAffiliation":null
      },
      "BestPaperAwarded":false
   }
}
```

In this representation, each property is encoded as a JSON name–value pair. Metadata for the entity appear under the __metadata name. By default, navigation properties are *deferred,* and each related entity is represented through its URI; when navigation properties are expanded, instead, each related entity is represented inline.

If a requested resource does not exist, the response is a 404 Not Found message. If it has been moved to another location, the response is a 303 See Other message. This message, specified in the Location response-header field, denotes the URI to visit to find the wanted resource.

To retrieve an individual property, a client issues a GET request to the property URI. The property URI is the entity request URI with "/" and the property name appended. The HTTP request

```
GET http://example.org/odata/Papers(5)/Name HTTP/1.1
```

returns

```
<Name>An Example Paper<Name>
```

In the case of a complex typed property it is possible to specify various granularity levels; for example,

```
GET http://example.org/odata/Papers(5)/Author HTTP/1.1
```

returns

```
     <data:Author metadata:type="Example.AuthorType">
           <data:AuthorName metadata:type="Edm.String">Alice
Smith</data:AuthorName>
           <data:AuthorAffiliation metadata:type="Edm.String"
metadata:null="true"/>
      </data:Author>
```

(the `metadata:null` attribute indicates a null value for the property corresponding to the element where it appears); whereas

```
GET http://example.org/odata/Papers(5)/Author/AuthorName HTTP/1.1
```

returns

```
<data:AuthorName>Alice Smith</data:AuthorName>
```

OData Collections

Retrieving a collection of entities, or a whole entity set, is similar to retrieving a single entity except that no key is specified:

```
GET http://example.org/odata/Papers HTTP/1.1
```

The entity set is mapped into a whole Atom feed, and each entity is represented as a feed entry.

```
<?xml version="1.0" encoding="utf-8" ?>
<feed xml:base="http://example.org/odata/v3/"
xmlns:data="http://docs.oasis-open.org/odata/ns/dataservices"
xmlns:metadata="http://docs.oasis-open.org/odata/ns/dataservices/
metadata"
xmlns="http://www.w3.org/2005/Atom">
  <title type="text">Papers</title>
  <id>http://example.org/odata/v3/Papers</id>
  <updated>2012-11-06T10:50:17Z</updated>
  <link rel="self" title="Papers" href="Papers" />
  <entry>
    <id>http://example.org/odata/v3/Papers(5)</id>
    <title type="text">An Example Paper</title>
    <summary type="html">
```

```
This paper contains examples of OData query requests.
    </summary>
    <updated>2012-01-02T11:12:00Z</updated>
    <author>
      <name />
    </author>
    <link rel="edit" title="Paper" href="Papers(5)" />
    <link rel="http://docs.oasis-open.org/odata/ns/related/
Review" type="application/atom+xml;type=entry" title="Review"
href="Papers(5)/Review"/>
    <category term="Example.Paper" scheme="http://docs.oasis-
open.org/odata/ns/scheme"/>
    <content type="application/xml">
      <metadata:properties>
        <data:Id m:type="Edm.Int32">5</data:Id>
        <data:Year metadata:type="Edm.Int16">2008</data:Year>
        <data:Name metadata:type="Edm.String">An Example Paper</
data:Name>
        <data:Author metadata:type="Example.AuthorType">
           <data:AuthorName metadata:type="Edm.String">Alice
Smith</data:AuthorName>
           <data:AuthorAffiliation metadata:type="Edm.String"
metadata:null="true"/>
        </data:Author>
      <data:BestPaperAwarded m:type="Edm.Boolean">false</
data:BestPaperAwarded>
      </metadata:properties>
    </content>
 </entry>
 <entry>
  ...
 </entry>
  ...
 <link rel="next" href="http://example.org/odata/v3/
Papers$skip=10" />
</feed>
```


The id element specifies the globally unique identifier assigned to the feed. The feed also contains a link element with the rel attribute set to "self." Following the URI in the href attribute (which is relative to the base URI set in the xml:base attribute), it is possible to retrieve the feed itself.

For optimization purposes, the number of entries contained in the feed may be less than the entities in the entity set. In this case, the feed contains a link element with a rel attribute set to "next" indicating the presence of additional entries. It is possible to follow the URI specified into the href attribute to fetch the next set of results. In the case of copious results, it is possible to filter, sort, page, and select a subset of entries to be returned. To this end, system query options are used. System query options have the form of a dollar sign followed by a specific command and are directly appended to the request URI. Table 7.1 illustrates some system query options and their meanings. The request

```
GET http://example.org/odata/Papers?$filter=Year%20gt%20
1998&$top=3 HTTP/1.1
```

returns only the first three articles in the Papers entity set that have their property Year set to a value greater than 1998. Note how the filter expression is encoded inside the URI. (The comparison operator greater than is encoded as *gt* and spaces are percent encoded as %20.) The filter expression supports logical and arithmetic operators, functions, and groupings. OData specifications contain more details on the set of allowed operators and built-in functions.

Handling Entities

Entities within a set can be created, updated, and deleted. An entity is created by sending a HTTP POST request specifying the URI of the entity set or collection where the new entity should be created. The body of the request message contains a representation of the entity to be created.

To update an entity, the HTTP PUT method may be used. The request URI must be set to the URI specified in the "edit link." The request message body contains a representation of the entity according to an allowed OData format (e.g., OData Atom format or OData JSON format). Any entity key appearing in this representation is ignored—entity keys cannot be updated.

The PUT method is used for a full replacement; that is, it overwrites a resource with a complete new body. To support differential updates, the HTTP PATCH method, proposed in RFC 5789 (Dusseault and Snell 2010), and the new HTTP MERGE (proposed in the OData specifications) method may be used instead. In these methods, only properties defined in the request body are updated, while other properties are left unchanged. With PATCH, the value of each property is updated to the one specified in the request body exactly. This implies that complex typed properties must be accurately specified component by component. With MERGE, however, updates happen on a component basis, and complex typed properties need to be specified only partially.

The HTTP DELETE method is used to delete an existing entity. The request URI should match the URI assigned to the entity to be deleted.

OData allows custom operations (functions and actions) to be executed. Operations are advertised in the service metadata document unless they are available only on specific entity sets or even on single entities. In the latter case, they are directly advertised inside each representation of the entity set or (single entity) returned by the service.

Operations are defined in terms of their name, return type, and parameters and are invoked by the client issuing a corresponding HTTP request: a GET request for functions and a POST request for actions. In the case of a function, the parameters appear inline and in the case of an operation, the parameters are passed as a JSON object inside the request message body.

Similar to methods in object-oriented programming languages, bindable operations are the operations for which the first parameter (called the *binding parameter*) is the entity on which the operation is performed. The binding parameter appears directly inside the URI used to invoke the operation. For example, an action `SetAuthor()` may be defined on the entity of type Paper as a convenience method to set name and affiliation of the primary author. This action takes three parameters: an entity of type `Example.Paper` upon which it is executed (the binding parameter), the name of the author, and the author's affiliation. The following POST request invokes the action

```
POST http://example.org/odata/Papers(5)/SetAuthor HTTP/1.1
{name:"Alice Doe", affiliation:"Free University"}
```

The action may return a result (e.g., a representation of the updated entity) in a 200 OK HTTP Response message, but the client can override this behavior by specifying the value `return-no-content` in the Accept request-header field. In this case a 204 No Content HTTP Response message is returned.

As a further example, a function `FavoriteByUser()` may be defined. This function is characterized by two parameters: a collection of entities of type `Example.Paper` (the binding parameter) and a username. To invoke the function, a client issues the following request:

```
GET http://example.org/odata/Papers/
FavoriteByUser(user=%34Bob%34) HTTP/1.1
```

and favorite papers are returned as an Atom feed.

SECTION III
Content-Centric Networks

Internet principles hinge on a paradigm derived from the inception of computer networks, host-to-host communication.* The Transmission Control Protocol (TCP)/Internet Protocol (IP) architecture behind the Internet was developed to ensure communication between two machines only one willing to supply a resource and one accessing that resource. Through its architecture the Internet protocol suite has created a strong coupling between the identity of a resource and its host in the network. Skyrocketing in recent years, Internet usage has created a constant need for more IP addresses, more hosts, and more servers.

In addition, the declining costs of data storage as well as connection have permitted access to a staggering amount of information (Gantz and Reinsel 2011 report that about 1.8 zettabytes of new information are created each year), demonstrating that the Internet is mostly a vehicle of content sharing (where *content* is the generic term used to identify any information made available to users) on the network. Van Jacobson recently observed that today people value the Internet for what content it contains, but communication is still in terms of where (Zhang et al. 2010).

This gap between content and communication noted by Jacobson is evident in a number of features users encounter daily. Network endpoints are assigned numeric IP addresses, but since the introduction

* Driven by Dave Clark's end-to-end principle (Saltzer, Reed, and Clark 1984), core networks are today relatively simple, and all the necessary intelligence is outside the network, into hosts.

of the Domain Name System (DNS) they are usually referred to by name. Built on domain names, Hypertext Transfer Protocol (HTTP) Uniform Resource Identifiers (URIs) virtually provide *handles* to all content on the web. All users need to do is type this handle into their browser's address bar. Their expectation is to access content by name, but browsers actually access locations, not content, so if the content changes location it must also change its name. One solution to this problem is to imitate name persistence, making content displacement transparent (e.g., using HTTP redirection).

Additionally, HTTP Secure (HTTPS) is mostly intended to secure the network and the communication channel but also to authenticate the identity of the remote communication party or website. In HTTPS, the client typically owns a certificate storage, used to determine the server's identity. This does not, however, guarantee the reliability of the information downloaded from the server (at least not automatically), which naïve users may assume that it does.

These days, business is looking for a more efficient network infrastructure, while most research on content networks is concerned with revising the original Internet addressing mechanism. The underlying idea of content-centric networking is simple. Instead of assigning addresses to hosts, assign them to content and route messages based on those addresses. Named content networking is currently the aim of several research projects, including content-centric networking (CCN; Jacobson 2009), translating relaying internet architecture integrating active directories (TRIAD; Gritter and Cheriton 2001), and data-oriented network architecture (DONA; Koponen et al. 2007) in the United States and SAIL (Ahlgren et al. 2008), the Publish/Subscribe Internet Routing Paradigm (PSIRP; Tarkoma, Ain, and Visala 2009), and CONVERGENCE (Blefari-Melazzi 2012) in Europe, just to name a few.

Surprisingly, the shift in perspective necessary to build routing schemes around content names is less disruptive than one might expect. This new design benefits from lessons learned from years of IP routing, forwarding, and congestion control algorithms. Evolving content networks are designed to exploit the same algorithms and even the same protocols used in the TCP/IP stack. Content networks may also incorporate important features typical of the Representational State Transfer (REST) architectural style, such as "smart" caching mechanisms and the possibility of requesting certain content even before that content exists.

Merging features from the network layer and the application layer may allow novel content networks to combine the advantages of an infrastructure designed for efficient delivery with content awareness resulting in success where traditionally IP multicast groups have failed.

8

Content-Oriented Communication and Conversational Communication

In general, today's Internet can be seen as a mix of content-oriented communication and conversation-oriented communication. A *content-oriented* communication pattern takes place when an entity in the network needs specific content irrespective of where it is stored. Content is associated with an identifier (or *name* in certain contexts), ostensibly unique in a specific namespace.

This simple communication pattern can be described for the network layer with a couple of primitives that allow the user to put content into and get content from the network. This content can be statically stored in a server or streamed live (audio, video, or multimedia); that is, the requesting node can start playing out the content without waiting for the full content to be transferred. So a content-oriented communication pattern can be classified as a file transfer, file streaming, or live streaming. A corresponding aspect of this pattern is the character of the communication, which may be unicast or multicast, where content can be requested by more than one node at a time (Figure 8.1).

A *conversational* communication pattern takes place when a user agent wants to exchange information with another in a bidirectional way. This pattern obviously includes messaging and audio and video calls but also other typical web services. For example, many social networking sites require a continuous exchange of information from the client to the server. These types of applications are not employed simply to show or play content but also to exchange information between client and server and thus fit the conversation communication pattern.

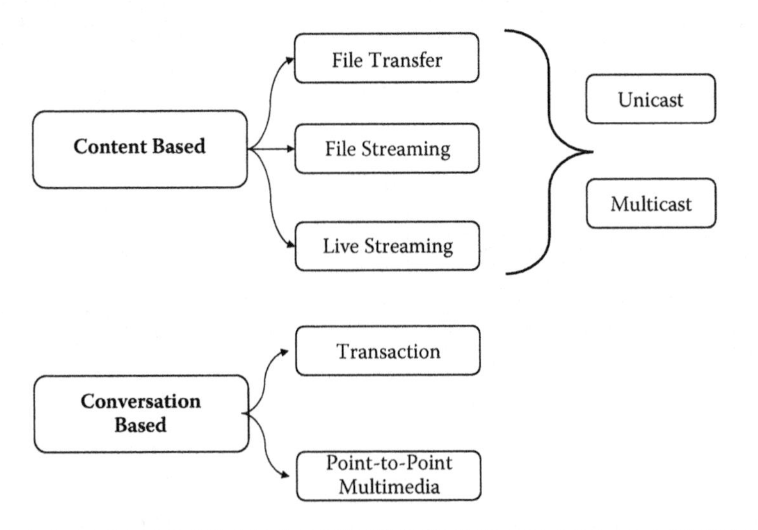

Figure 8.1 The problem space and the solution space of content-centric networking.

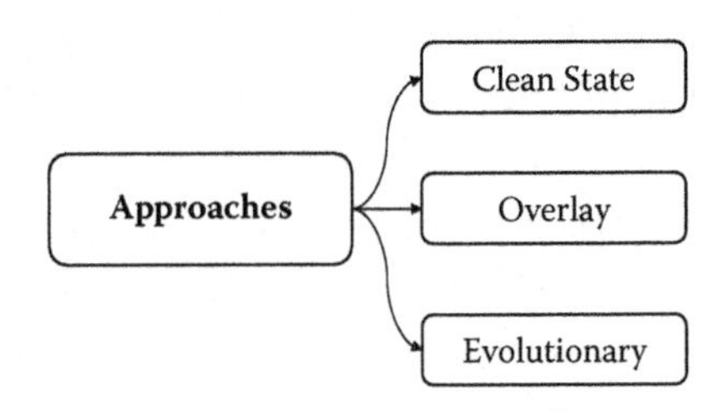

Figure 8.2 Clean slate, overlay, and evolutionary approaches to content-centric networking.

Clean Slate, Overlay, and Evolutionary Approaches

The shift from networks of hosts to content-centric networks is still largely a research topic. The literature takes three main approaches (Figure 8.2).

The *clean slate* design is the simplest approach. It aims to substitute a content-oriented layer that will sit directly over the data-link layer 2 (e.g., Ethernet) for the existing Internet Protocol (IP) layer. Conversational patterns can still be supported by copying them on top of the content-oriented mechanism (Jacobson et al. 2009), but the clean slate approach can hardly be applied in the near future because it requires the replacement of the entire IP network, which is impractical at best.

In the *overlay network* approach, units of information content are tunneled via a Transmission Control Protocol (TCP) or a User Datagram Protocol (UDP) over IP. This approach, typically implemented in peer-to-peer networks (Balakrishnan et al. 2003), is often referred to as a temporary "hack" that positions a content-centric network over the current network. The CCNx project, currently in development at the Palo Alto Research Center in California, is an example of an overlay solution that transports content in binary encoded Extensible Markup Language (XML) packets of variable lengths over the current IP technology.

A third overall method is the *evolutionary* approach where content-centric networks may operate within existing networks (Gritter and Cheriton 2001; Koponen et al. 2007). The evolutionary approach has the advantage of maintaining the current network infrastructure while allowing for coexistence of conversational and content-oriented communication patterns.

Naming Scheme

The selection of an accurate naming scheme will strongly impact a content-centric network in terms of scalability, usability, and security. Zooko's Triangle (Zooko 2006) illustrates the general problem of choosing the best naming scheme (Figure 8.3). The three properties highlighted in this diagram are uniqueness, human readability, and distribution.

All existing naming schemes share at most two of these three properties. For example, Domain Name Systems (DNSs) are memorable

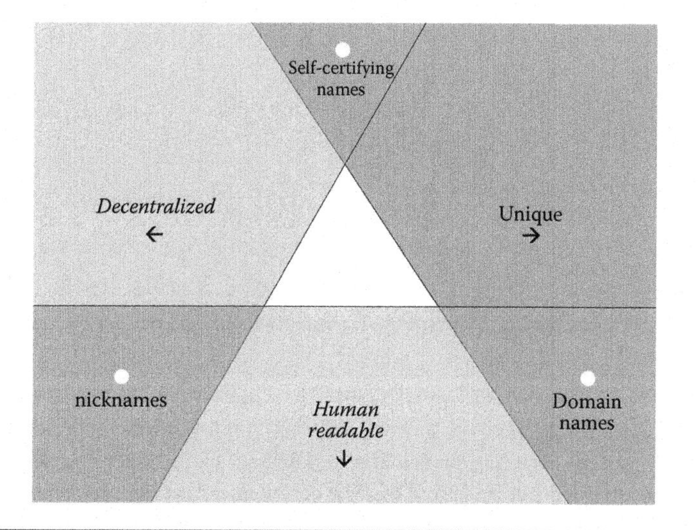

Figure 8.3 Zooko's Triangle.

and unique, but they need a centralized authority to be resolved. The nickname is human readable and decentralized, but by virtue of its nature cannot be unique. Data keys in peer-to-peer networks and self-certifying names* are unique and might be issued in a decentralized way but are not human readable. By properly combining two schemes, however, a system may achieve all the three properties (Stiegler 2005).

* Examples of self-certifying names include Digest such as SHA-1 applied over a whole content or part of it as proof that the content is authentic; names that include the identifier of a public key and a digital signature from the corresponding private key—e.g., in X.509 certificates.

9
Web Content Delivery Networks

In previous attempts to deliver content directly over the Internet Protocol (IP) multicast channels using, for example, class D IP addresses, a total lack of control of both the channel by content producers and the content by consumers is evidenced making this technique charming from an academic viewpoint but less appealing indeed from a business perspective. Nowadays content is mostly accessed on the web, but the web as a distributed system still depends on web servers, which are centralized systems subject to the weakness of a single point of failure, namely, the origin server. As a consequence of both the evolution of technology and Hypertext Transfer Protocol (HTTP) specifications, a wide range of content delivery solutions (Hofmann and Beaumont 2005) have emerged over the last few years that increases reliability.

One obvious solution to improve the availability of web content is to replace the origin server with a pool of origin servers and to exploit the features provided by the Domain Name System (DNS). Entering multiple Type A resource records specifying that the same host name can be resolved in a number of IP addresses allows a DNS server to resolve the same name into a number of different hosts, using a round-robin strategy. Because the time-to-live (TTL) field in the record indicates how long the latter can be kept valid in a nonauthoritative server's cache, the TTL could in principle be used to predict how long a client will rely on a particular origin server.*

A slightly more sophisticated approach involves the introduction of a web server farm. In the server farm model, an array of servers shares the burden of answering requests for the same website. All of the

* A similar approach is used by the DNS itself when root name servers have to be found by the hostname. The hostname of a root server is actually resolved into a pool of name server addresses.

servers are interchangeable, so their workload is reduced. Requests are routed to a load balancer before arriving at the servers because the load balancer is configured with the public IP address assumed by the client to be that of the origin server. Origin servers, by contrast, are typically not public but are configured with private IP addresses. The balancer may use a number of strategies to detect the server most appropriate to respond to the request.

Compared with simple name server resolution, this approach introduces the concept of load balancing. Some load balancers operate at the transport level as well to route requests to a specific server based on the Transmission Control Protocol (TCP) port number specified in the packet carrying the request. For example, they might route HTTP, HTTP Secure (HTTPS), and File Transfer Protocol (FTP) requests to three different physical servers. The load balancer maintains associations similar to the ones maintained by a Network Address Translation (NAT) between source clients and destination servers. These associations are usually per-source IP address (and port) but sometimes even per user and may be used to store, for example, information in HTTP cookies, thus allowing a server to sustain transactions. This feature is usually called *persistence.*

All the previous solutions work at the origin server level. Contrasting solutions such as caching works on an intermediate level between the client and the origin server. A shared cache (i.e., a cache shared by multiple clients) may be placed near a specific group of content consumers, near the origin servers that host the content, or somewhere inside the network at a halfway point (Figure 9.1).

A *forward proxy* represents the first kind of cache. The forward proxy cache operates by assuming that when several consumers belonging to the same workgroup request content, it is more than likely that they will request the same content which increases the probability of a cache hit and reduces the workload at the origin server.

The second kind of cache is any gateway or reverse proxy. A gateway, usually placed near the origin server (farm),* improves content delivery

* Sometimes different reverse proxies may be placed close to different groups of consumers instead of the origin servers. This solution is intended to combine the advantages of a forward proxy (in particular, the reduced latency in content provisioning for a group of clients) with those of a reverse proxy.

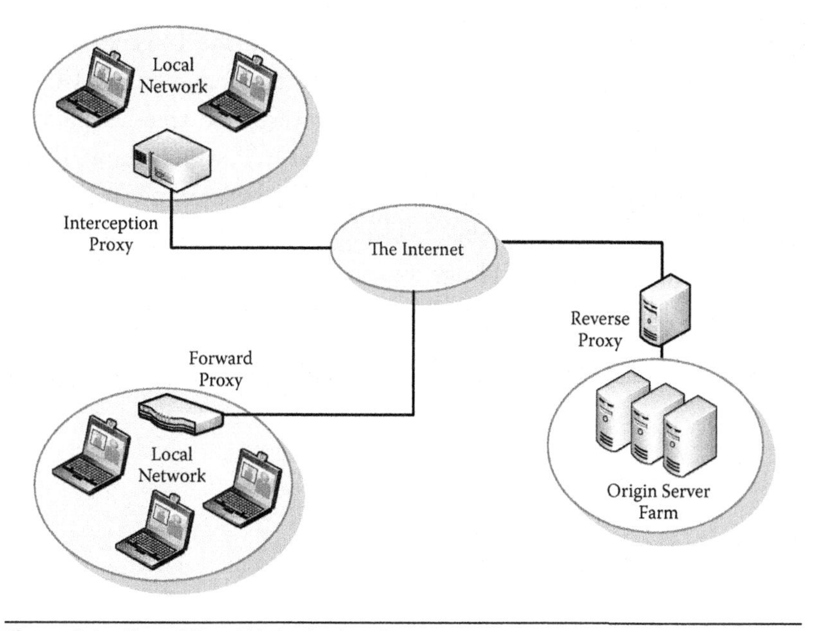

Figure 9.1 Three different kinds of web caches.

by reducing the burden of requests to the origin server, allowing it to focus on producing rather than serving content. The gateway is configured with the DNS address where the client would expect to find the origin server. In this way the gateway, not the origin server, receives requests from clients. Client requests are forwarded to the origin server if and only if the content is not found in the gateway's cache.

A third kind of shared cache may be placed inside the local network, using an interception proxy. This approach differs from a forward proxy in that it requires a dedicated appliance (web switch) on the local network to intercept web traffic (e.g., TCP connections on port 80) and route it to the interception proxy shared cache (instead of to the origin server outside the local network). The interception proxy is useful whenever the administrator cannot possibly configure a forward proxy for a client in the network of an Internet Service Provider, for example. The use of an interception proxy or any other "middlebox" in the network has often been criticized due to its violation of the principle of network transparency, which says that packets should flow from source to destination unaltered (RFC 2775, Carpenter 2000).

Web caches may be arranged in chains at different (usually geographic) levels. For example, there may be regional caches that are served by national ones, which in turn are served by origin servers.

However, a more efficient schema is defined in the Internet Cache Protocol (ICP; RFC 2186 and RFC 2187, Wessels and Claffy 1997a, 1997b), which arranges web caches into “forests and trees” routed in origin servers and defined by a protocol called ICP, which allows them to exchange User Datagram Protocol (UDP) messages to learn which has the specific requested object. Using ICP, each cache node may request content from its siblings and, in case of a miss, may refer to its parent. This procedure is repeated until an origin server is reached.

10
Peer-to-Peer Networks

Originally the Internet was a network of peer hosts. The progressive expansion of client–server protocols is just one development that has made this design less obvious. For example, transient Internet Protocol (IP) addresses assigned by Dynamic Host Configuration Protocol (DHCP), Serial Line Internet Protocol (SLIP), and Point-to-Point Protocol (PPP) contribute to the increase of hosts on the Internet, allowing virtually any computer in the world to become a host but also creating asymmetry between transient nodes and nodes with static IP addresses. Similarly, the ever increasing number of local intranets with firewalls and Node Auto Terms (NATs) preventing nodes from being contacted by the Internet as well as local domain name servers assigning host names not visible from outside the intranet have contributed to this asymmetry.

RFC 2775 (Carpenter 2000) enumerates several other reasons that today's Internet has lost its symmetrical nature to become a network based on centralized hierarchical agreements.

Peer-to-peer networks tend to reclaim the original symmetry and to reenable the original idea of a direct exchange of data between hosts participating in the peer network. Many implementations of peer-to-peer networks have emerged, providing services such as content sharing, distributed storage, and distributed computation.

Unlike client–server networks, peer-to-peer networks share resource ownership across all peers. In principle, this implies an increased reliability as there is no single point of failure and peers rely on each other for service. Furthermore, such a network is intrinsically dynamic due to hosts' joining and leaving relatively frequently. These features mark the obvious difference with traditional network environments and require different protocols, algorithms, and routing strategies.

Napster

Indeed, the primary problem to solve in a peer-to-peer context is lookup. Early peer-to-peer systems mainly focused on handling large-granularity objects, typically "opaque" files identified by their name.

Napster, probably the first and most widely known peer-to-peer network, adopted a centralized database in 1999 to solve this problem. Napster's database maintains an index that maps a content name to an IP address of nodes that host the content. A Napster node uses a proprietary Transmission Control Protocol (TCP)–based protocol to contact the Napster server and retrieve location information. It then connects directly to a peer host to download the content. This centralized index approach has scalability and reliability problems, however, and requires a powerful infrastructure that results in vulnerability of the client–server architecture, thus eliminating the benefits of a peer-to-peer system.

Gnutella

The Gnutella network adopts an approach based on broadcasting. Gnutella "servants" join the Gnutella network by contacting *host cache servers*, prominent network nodes whose function is to cache the IP address of the contacting node. This approach both provides the joining node with a list of servants already in the network and caches the address of the new node, which is likely to become a servant. Once the new node is connected to at least one node of the Gnutella network, it can forward discovery messages to find additional nodes and establish connections with them, using a TCP-based protocol. Outgoing messages contain a time-to-live (TTL) field. Peers receiving the message forward it to other peers with whom they are connected (avoiding loops). The TTL is decreased at each hop. Messages whose TTL is zero are discarded to avoid flooding the network. Table 10.1 illustrates the different kinds of messages used in Gnutella. Ping and pong messages are used to discover other peers, whereas query and query hit messages lookup content. Once content is found, the requesting node establishes a Hypertext Transfer Protocol (HTTP) connection to the servant where the content is located and downloads the file. Download of one or more subranges (in terms of bytes) of the

Table 10.1 Main Different Types of Messages in Gnutella Network

MESSAGE	SUMMARY
Ping	This message is issued by a peer willing to connect to another. The node is asking another peer to respond with a pong message.
Pong	Response from a peer node willing to accept an incoming TCP connection. Contains the IP address and TCP port with which to connect.
Query	A query that contains a description of the requested content.
Query hit	Positive answer to a query message. Contains the IP address and TCP port with which to connect. The requesting peer issues a HTTP request at that IP and port to download the content.
Push	Instructs the servant to establish a connection with the requesting node instead of receiving an incoming connection. Useful when a servant is behind a firewall and cannot receive incoming connection.

requested resource instead of the whole content can be interrupted and resumed using the HTTP Range request-header field.

The drawback of the Gnutella network is that the servant discovery strategy implemented through a ping message broadcast to all peers is inefficient. It has been confirmed that about two thirds of all messages (Zeinalipour-Yazti and Folias 2002) are needed to discover peers and only a small fraction are query hits containing indications of where the requested content is located. Servant discovery messages contribute to the increase of the network load without providing any effective benefit to a peer.

To improve lookup and reduce network traffic, Gnutella introduced ultra-peers. Each ultra-peer is connected to more than 32 other ultra-peers, whereas other nodes or "leaf nodes" are connected to only a small number of ultra-peers. Ultra-peers modify the topology of the network in a hierarchical fashion. In fact, leaf nodes exploit ultra-node connections to publish and search for content instead of connecting directly to other peers. This allows it to decrease the TTL of each message, improving performance and reducing network traffic.

The ultra-peer approach has been adopted in more recent peer-to-peer networks based on the FastTrack protocol* where ultra-peers are called supernodes. It requires specialized nodes with more powerful hardware platforms and higher bandwidth connections and thus makes the peer-to-peer network asymmetric.

* File sharing programs using FastTrack include Kazaa, Grokster, iMesh, and Morpheus.

11
Distributed Hash Table

Distributed hash tables (DHTs) are an alternative to traditional lookup systems. They are used in symmetric distributed lookup algorithms where lookup is performed by following references that the various nodes maintain until an appropriate node containing the wanted data (or a duplicate) is found. The attribute *symmetric* means that all nodes are peers and perform identical work. DHTs arise from the necessity to improve reliability and distribution of hierarchical lookup systems such as the Domain Name System (DNS) and similar peer-to-peer lookup systems. In the DNS, the hierarchical approach leads to the need for well-known root name servers and top-level domain (TLD) name servers that must meet the burden of countless requests daily. The workload of these servers is typically reduced by the local resolver's cache and by caching name servers, but obviously this is acceptable only in a relatively static context, not in a highly dynamic content network.

DHT-based interfaces have recently been adopted as scalable and reliable solutions in a number of peer-to-peer systems. They may be seen as an improvement over previous hierarchical solutions (e.g., FastTrack), but their primary application today is in distributed storage systems. DHTs are mainly used to implement exact-match lookups; Harren, Joseph, and Huebsch (2002) discuss the feasibility of supporting more complex queries using algorithms based on DHTs.

In DHT-based lookup systems, each content and each node is associated with a numeric key respectively called a *data key* and a *node key*. The key is obtained by applying a hashing function to the content name (which is assumed to be unique) or the node identifier (e.g., its Internet Protocol [IP] address). Nodes have assigned keys in the same key space as content. A content is assigned to a node when its node key is close (in response to an ordering algorithm) and, in turn, assigned to the data key associated with that content. The chosen ordering algorithm is often the simple, natural numerical order. For instance, consider a hash function that returns keys in the range [0,255]. A network has three nodes, say 10.0.0.4, 10.0.0.7, and 10.0.10.3, whose node keys are, respectively, 012, 065, and 200. Consider then content A's name hashed into 014. Content A will be associated to node 10.0.0.4 because in the increasing natural numerical order that node is considered the closest because its node key is less than 1000. Similarly, content B, which is hashed into 211, will be given to node 10.0.10.3. A content that is hashed into a value lesser than 012, say 005, would be circularly assigned to node 10.0.10.3.

The previous example also illustrates that to achieve an effective load balancing it is important to have node identifiers uniformly distributed across the whole range of keys. This is generally achieved by assigning each node a random identifier within the key space.

Nodes are required to implement that DHT interface, that is, to provide a lookup method that, given a key as input, returns a reference to the next node to lookup. The algorithm does not mandate the format of the reference though typically DHT algorithms are implemented on overlay networks. Thus, a reference could be in the form of an IP address (as in the previous example), a domain name, a Uniform Resource Indicator (URI), or whatever identifier is given to the nodes in the network. The transport is not part of the algorithm and is typically left to each implementation. However, routing is part of the algorithm. Each node contains a routing table with relatively few entries. This routing table is used to forward, hop by hop, a lookup request to a node whose key is closer to the key contained in the request.

Chord

Nodes in Chord (Stoica et al. 2001) are arranged into a ring whose length, D, corresponds to the extension of the key space. Each node has associated its *node key*, say k, and maintains a "finger table" containing the IP addresses of nodes distributed across the entire key range. Entries in the finger table follow a logarithmic distance rule so the first entry contains a node key assigned to a node that is located approximately over the opposite position (halfway around the ring). To the latter are forwarded requests for keys greater than n + D/2 (modulus D) and lesser than n + D/2 + D/2 = n + D (modulus D). The second entry is a node that is over a quarter of the way from n (if it exists) and is responsible for handling requests for keys greater than n + D/4 (modulus D) and less than n + D/4 + D/4 = n + D/2 (modulus D) and so forth.

Following its finger table, each node routes lookup requests to a node that handles half of the initial node's key space portion so that the search converges in logarithmic time. Also, this strategy allows the number of hops to scale as a logarithmic function of the number of nodes.

Figure 11.1 illustrates how key 57 can be found on a ring with C = 64. The query starts at node 0. Because 57 is farther from 0 than C/2 = 32, the node forwards the query message to a node that is approximately in the opposite position (node 38 in this example). Node 38 proceeds with the same test and finds that the best node to send the query to is about a quarter of the way around. So it forwards the message to node 48. The message finally comes to node 53, which is responsible for storing the requested key. The whole process converges in logarithmic time.

A new node joining the network receives a randomly assigned node key. The joining node then sends a particular lookup request to an arbitrary node already connected to the network. The lookup

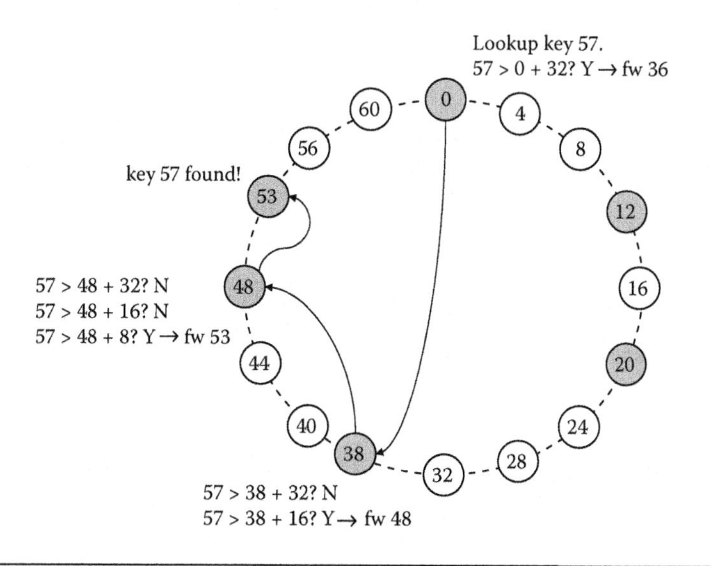

Figure 11.1 Node 0 locks up for key 57 in a Chord network.

request contains the node key. Upon receiving the request, nodes that are predecessors of the joining node (i.e., nodes whose node keys are less than the new node key) update their successor lists. This ensures that lookup requests are still correctly met while the joining node and existing nodes update their finger tables.

Because nodes may fail or leave the network, each node in Chord maintains a list of numbers of its successors in the ring. This ensures that a query message is forwarded along the ring even if some finger table entries are no longer valid because of a node failure.

Pastry

Pastry (Rowstron and Druschel 2001) uses a routing and forwarding mechanism very similar to Chord, but instead of searching for a node whose key is numerically close to the requested data key, messages are forwarded to a node whose key has more matching digits. A lookup message progresses around the ring one digit per hop, converging in logarithmic time. Kademlia (Maymounkov and Mazieres 2002) implements a similar strategy as well.

For example, assuming the key space is base 16, each peer maintains a set of routing vectors each made of 24 – 1 = 15 entries. Each entry contains the IP address of nodes whose key shares the first p (prefix) digits and differs from the p + 1 subsequent digits. If the key space contains C different values, then the number of different digits in a key is $n = \log_{16}C$. This way each node maintains exactly n routing vectors that form a bidimensional routing table. An example is outlined in Table 11.1, which shows a possible routing table for the node

Table 11.1 An Example of Pastry Routing Table at Node 05A3C0

—	1* ?	2* 10.0.0.5	3* 10.0.10.2	4* 10.12.5.5	5* ?	...	F* 10.0.12.3
00* ?	01* 10.1.9.7	02* 10.0.5.5	03* 10.0.0.7	04* ?	—	...	0F* ?
050* ?	051* 10.0.0.27	052* ?	053* 10.3.4.1	054* 10.0.0.25	055* ?	...	05F* 10.0.1.20
05A0* ?	05A1* ?	05A2* 10.1.5.18	—	05A4* ?	05A5* 10.0.0.15	...	05AF* 10.0.0.55
...	...	...	...	...	...	...	...
—	05A3C1* 10.0.3.1	05A3C2* 10.0.3.16	05A3C3* ?	05A3C4* 10.0.0.67	05A3C5* 10.0.1.66	...	05A3CF* ?

with key 05A3C0. Each cell contains a key prefix and may contain the corresponding IP address of a node, if known, whose key matches the prefix. The length of the prefix increases with the number of rows. Empty cells in each row correspond to prefixes matching the node key of the node hosting the table. Cells marked with a question mark (?) in the table are not associated to a node. Additionally, each node maintains a *leaf set*, that is, a list of nodes that are its immediate successors and predecessors.

Upon receiving a message, a node scans its leaf set to learn if the requested data key is covered by one of the nodes in that list. If there is a match, the node forwards the message to that node. If not, the node uses its routing table to find the best node (the one whose key has the longest prefix match) to route the message. The example covered in Table 11.1 illustrates that upon receiving a query for data key 0534FA the node associated with node key 05A3C0 forwards the message to the node in its routing table that has the longest prefix matching 0534FA (i.e., 053*), which corresponds to the IP address 10.3.4.1. If no node with a higher number of "correct" digits is found, then the node forwards the message on to any other node with a node key sharing its same key prefix but numerically closer to the requested data key.

12
JXTA Project

Historically, JXTA has been one of the most relevant efforts to build peer-to-peer platforms. Project JXTA was thought to allow any network device to exchange information without a centralized infrastructure. JXTA specifications define a common set of protocols for building overlay peer-to-peer networks. The protocols are based on Extensible Markup Language (XML), and thus they are independent from any language, operating system, and network. Implementations of JXTA specifications exist in many modern computer languages and platforms. JXTA 1.0 protocol specifications were released in April 2001, and version 2.0 was available in late 2003.

JXTA is an overlay network. The protocol stack is arranged in logical layers on top of the transport layer (Figure 12.1). Commonly used transport protocols include Transmission Control Protocol (TCP), User Datagram Protocol (UDP), logical transport protocols such as Hypertext Transfer Protocol (HTTP), and their secure variants. JXTA is not, however, limited to Internet Protocol (IP) networks, and any peer may use other transport protocols specific to the network to which the peer is connected (e.g., a personal area network using Bluetooth).

Each resource in JXTA is associated with a unique identifier called a JXTA Identifier (JXTA ID). The JXTA ID is a Uniform Resource Name (URN). As opposed to using a locator such as a Uniform Resource Locator (URL), this option allows each peer to change its localization address while still keeping its URN. It is particularly suitable for nomadic peers.

Each ID is composed of three parts: a namespace identifier (jxta), a format specifier, and a value. The format specifier is extensible and any application could define its own, but a common choice for resources is a universally unique identifier (UUID), that is, a 128-random-bit sequence generated at the creation of the resource assigned to that and

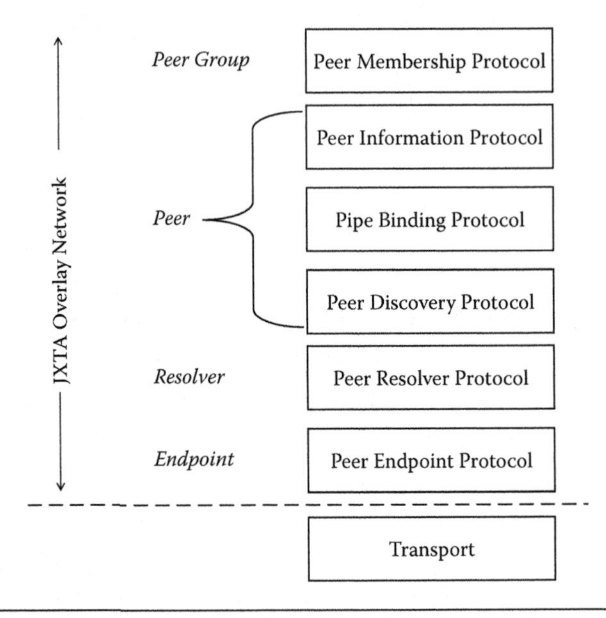

Figure 12.1 JXTA architecture.

only that identifier. For example, a JXTA peer could have the following JXTA ID:

```
urn:jxta:uuid-<128 bit sequence>
```

Another format specifier, jxta, is used to define reserved IDs. Among reserved IDs there are the NULL ID and the Net Peer Group ID. The Net Peer Group is the group to which all peers belong by default when they first connect to the JXTA network.

Peers

Peers are classified into (1) simple peers called *edge* peers to highlight that they do not perform particular functions for the network but stay at its edge, (2) super-peers, (3) rendezvous peers, and (4) relay peers, which instead provide lookup facility and traversal services, respectively.

Peers connect to other peers through pipes that attach to their endpoints. An endpoint is an abstraction that encapsulates all physical network addresses associated with a peer. There is one endpoint for each peer participating in the JXTA network, and within that endpoint each peer can dynamically select the most efficient network interface to communicate with another. The Peer Information

Protocol (PIP) is a simple protocol to monitor peers' status. The peer receiving a PIP query message may reply by providing information about its uptime, generated network traffic, or other parameters.

Pipes

Pipes are defined as virtual asynchronous communication channels. Each pipe is unidirectional and attached to one peer endpoint. A pipe is called an input pipe or output pipe depending on whether the peer uses it to send or receive messages. To establish a communication channel, one output pipe from one peer should be coupled with the input pipe from another (simple unicast pipe). There are also secure unicast pipes, which guarantee reliable message delivery through acknowledgments, and propagate pipes, which are able to act on input or output for more than one peer (Figure 12.2) though they are detached from other pipes. The Pipe Binding Protocol (PBP) allows peers to find appropriate endpoints to attach to a given pipe and to reply to pipe binding requests by other peers, thus creating a virtual communication channel between two or more peers.

Pipes allow for the delivery of messages from one peer to another. Messages are XML based to provide maximum compatibility with different operating systems and programming languages and are

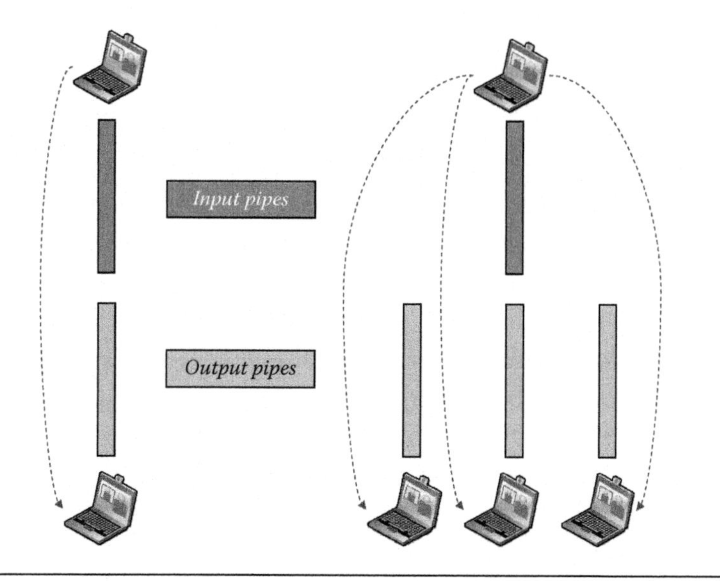

Figure 12.2 Different kinds of pipes.

composed of a list of key–value pairs, where the value is a typed content. Messages also contain opaque tokens, called credentials, that are used for authentication (and for authorization in some cases). The format of a token is not specified but is left to the specific implementation.

All JXTA protocols are asynchronous and based on the query request–response paradigm. The Peer Resolver Protocol (PRP) provides an abstraction for encapsulating all query request–response messages. PRP authorizes each peer register handler for the types of messages they are interested in receiving. This protocol takes care of dispatching outgoing query requests and receiving corresponding responses that then pass to the appropriate registered handler. The resolver encapsulates all the resolution operations needed at the underlying transport level (e.g., resolving a name to an IP address, binding a socket to a port, locating a server via a directory service). Additionally, the resolver performs verification of credentials in messages, discarding invalid messages.

Peer Groups

Peer groups are groups of peers. Each peer may belong to more than one group. Groups may have an open membership policy or require identification of each member peer. The Peer Membership Protocol (PMP) is used to authenticate the peer and to prove that the peer is a valid member of the group.

Each group typically implements its own collection of specific services called the *peer group services* to facilitate common operations, but each group may implement a number of standardized *core peer group services* (e.g., discovery of group resources, authentication, authorization, member communication, query, monitoring) as well.

In general, peers with the same interests will tend to join the same group. Forming groups is encouraged as this allows for the creation of different virtual secure broadcast domains.

By default, at the bootstrap every peer joins a default group called a *net peer group*. To discover other peers or other resources in general, a peer may use the Peer Discovery Protocol (PDP). The net peer group supports the PDP, whereas other groups may use different or proprietary discovery protocols. When using the PDP, the peer issues a discovery query specifying which kind of resource it is looking for and a maximum number of entries to be returned. The peer may send this query to other peers in the group (JXTA 1.0 only) or to a rendezvous peer—suggested in JXTA 1.0 and mandatory in JXTA 2.0—and may receive zero, one, or more responses, each of them containing an advertisement matching the query.

Modules and Codats

Module classes and module specifications describe functionalities provided by peers and peer groups. Each module specification describes an abstract functionality, that is, the *interface* of a service. Module specifications are implemented by one or more concrete module implementations. Module specifications describing the same functionality are grouped into classes. A codat is a content of any type identified by its JXTA ID. Codats are shared among members of the same group and cannot stay in more than one group; the same content shared in different groups would be associated with two different IDs, thus representing two codats. Codats can be replicated (copied) into several peers of the same group to improve availability.

Advertisements and Rendezvous Peers

Each resource is represented in the JXTA network through advertisements with XML documents describing the resource. Advertisements are published by peers and discovered by other peers. Advertisements expire and need to be republished if the resources they describe endure past the advertisement's expiration. An advertisement may be cached in the local memory of the peer. This mechanism is useful as it provides endurance and avoids having the peer look for new resources each time it reconnects to the network. When the advertisement expires, however, the peer needs to query other peers on the same local network or to use a rendezvous peer.

To facilitate resource discovery, JXTA specifications define superpeers called rendezvous peers. The rendezvous peer maintains a list of served edge peers and a list of other rendezvous peers to which it forwards queries that cannot be locally answered. A collection of rendezvous peers that are known to each other form a *propagation network*. In JXTA 1.0, a rendezvous peer consists of a repository where advertisements from peers may be stored. A peer searching for a particular resource can query the rendezvous peer. The RendezVous Protocol (RVP) ensures that messages in the propagation network do not loop and that their propagation concludes in a finite time.

JXTA 2.0 rendezvous peers do not cache advertisements anymore, but they maintain an index (the Shared-Resource Distributed Index, SRDI) of advertisements published by their served peer. The SRDI is based on a loosely consistent distributed hash table (DHT). Each rendezvous peer maintains a partial knowledge (rendezvous peer view) of the other rendezvous peers participating in the DHT. Whenever an edge peer needs to publish new advertisements, it pushes the advertisement's hash key to its rendezvous peer. The rendezvous peer maps the key to a target rendezvous node appearing in its view to handle that key.

As opposed to other DHT-based networks, JXTA does not guarantee the consistency of all peer views because of the high maintenance cost this would require in a highly dynamic environment where peers may join and leave frequently. Each rendezvous may have a temporary or even a permanent inconsistent view. To increase the hit ratio despite inconsistency, the index is also replicated on rendezvous nodes in proximity to the target node, that is, to rendezvous serving an adjacent range of keys in the DHT ring. When receiving a query from an edge peer, the edge peer serving rendezvous computes the DHT function to find the wanted hash key and, in turn, the target rendezvous node responsible to handle that key. If this target node has disappeared, the query is forwarded to nodes in its proximity because these store a replication of the key. If there is a hit, the query is forwarded to the peer that has published the wanted advertisement. This peer eventually replies to the requester.

If rendezvous nodes in proximity have disappeared, a *limited-range walker* strategy is used. The query begins to "walk" around nodes in proximity of the target node for a certain number of hops trying to find a hit. The walk is bidirectional, toward rendezvous nodes handling a range of keys greater than and less than the wanted key. The walk is stopped in each direction of the DHT ring when one of the following events occurs:

- A hit is found.
- Its time to live expires.
- There are no more nodes to walk in that direction.

Proposals have been made to improve RVP using the original Chord algorithm instead of the limited-range walker strategy (Nocentini, Crescenzi, and Lanzi, 2009).

Message Routing and Forwarding

To traverse firewalls and Node Auto Terms (NATs), JXTA specifications introduce the relay peers. A *relay peer* is a particular peer able to accept incoming connections from peers behind a firewall or NAT and to route these messages to the destination peer, supposedly located outside the local network. The relay peer (Figure 12.3) also serves the opposite purpose in that it accepts incoming messages from peers outside the local network and stores them until the destination peer, supposedly inside the network, connects to get the message.

In a dynamic network such as JXTA, where peers may join or leave at any time, static routing tables are not enough. JXTA specifies an ad hoc protocol, the Peer Endpoint Protocol (PEP), which takes care of message routing and the handling of firewall and NAT traversal.

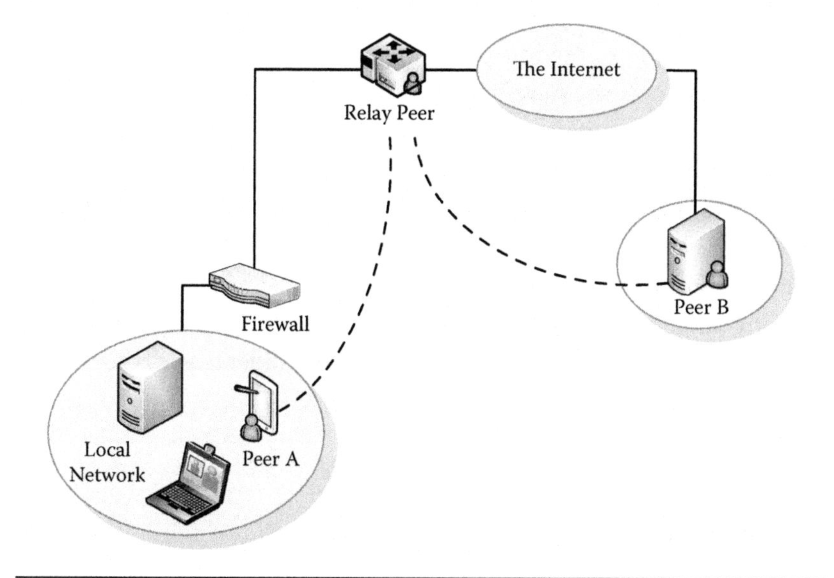

Figure 12.3 JXTA relay peer.

This protocol sits directly on the transport layer. Each peer maintains a local routing table with a limited number of reachable peers, including any relay peers, and the corresponding route advertisements. When the sender peer needs to send a message, it first looks for the message's destination in its routing table. If no entry is found, the peer sends a route query request to the local relay peer. The relay peer looks in its routing table, and if a match is found it sends back a response containing the route advertisement needed to reach the destination. If not, the relay peer contacts other relay peers recursively until it knows a route has been found.

When the route advertisement comes back to the sender, it is cached in the local memory for later retrieval and then used to send the message to the destination.

In any outgoing message the sender includes the forward route and the corresponding reverse route (or at least the route it believes to be the reverse). Any forwarded message also contains a *last hop* field, matching the last peer transmitting the message. At each hop the peer receiving the message inspects the destination. If it is the destination, no forwarding is necessary. If not, the peer checks the message for any loop. If the peer is the sender of the message or it matches the last hop, then there is a loop and the packet is discarded. The peer then checks it to see if the destination reported in the message is directly reachable. If this is the case it changes the forward route by removing the unnecessary hops and leaving only the destination as the next hop. If not, it checks the next hop of the route, and if it is not reachable it asks its local relay peer to find a new route. If no route is found, the packet is simply discarded.

The peer then checks the reverse route. It verifies that the last hop is directly reachable, and if this is the case it adds (if not already present) that peer to its routing table. If not, it asks its local relay peer for a route to that peer and updates the reverse route contained in the message.

The peer finally updates the last hop with its identity and forwards the message to the next hop.

13

Named Data Networking Project

Named content networking was first proposed in Jacobson, Smetters, and Thornton (2009), subsequently attracting several related contributors working on projects that were eventually integrated into a more general effort known as the Named Data Networking (NDN) project (Zhang et al. 2010).

The NDN architecture exploits the advantage of a universal central layer with a simple communication interface and is illustrated by an hourglass model (Figure 13.1) intentionally similar to the one proposed for the Transmission Control Protocol (TCP)/Internet Protocol (IP) stack (Deering 2001).

In both stacks, each layer represents an agreement between parts of the communication operation. For instance, layer 2 (the framing protocol) is an agreement between the two ends of a physical link. Layer 4 (the transport protocol) is an agreement between a producer and consumer. There is only one layer that requires a universal agreement: the network layer (i.e., IP). Much of the IP's success is due to the simplicity of the network layer, to its "thin waist" in the hourglass model, and to the lack of demand the network makes on its underlying layer, a stateless, unreliable, unordered, and best-effort service-delivering frame.

NDN architecture borrows a number of proprieties from the TCP/IP suite. For example, routing is decoupled from forwarding but operates in tandem (packets are forwarded while routing tables continue to evolve over time) the same as in current IP networks. At the same time, however, NDN adds specific new features. Three of these are seen as new layers in the NDN protocol stack. The strategy layer, which allows NDN to take advantage of multiple simultaneous connections (e.g., Ethernet plus 3G plus Bluetooth plus 802.11), is designed to use granular dynamic optimization choices to best exploit multiple connections

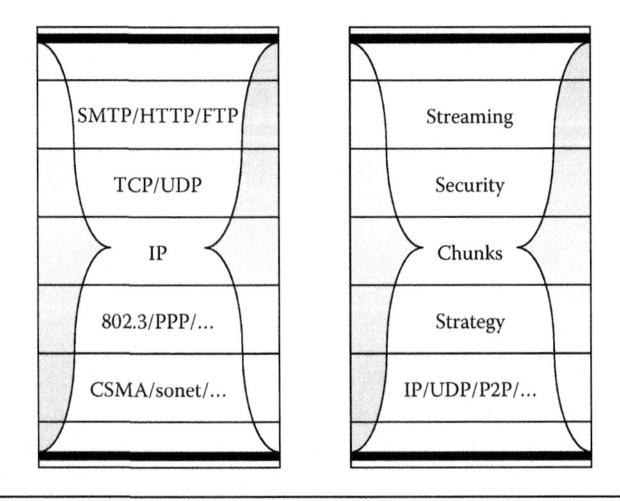

Figure 13.1 IP and NDN hourglass models.

under changing conditions. Layer 4 incorporates new proprieties such as self-regulation of network traffic (instead of relying on flow balancing at the transport level) and the possibility of the NDN level to be layered over anything (even the same IP protocol).

NDN provides a basic security building block right at the thin waist of the stack by signing all named data. It also applies security to content itself rather than making the connection on which the content is transmitted secure, thereby avoiding many of the host-based vulnerabilities that plague today's IP networks.

How NDN Works

NDN communication is receiver driven. A consumer willing to receive data sends out an Interest packet, specifying the name (identifier) of the desired content.

A router records the request interface then forwards it according to its Forwarding Interest Base (FIB), which is a table storing routes populated by a name-based routing protocol.

When the Interest packet reaches a node that "has" (but not necessarily the one that generated) the requested data, a Data packet is sent back. A Data packet carries the name of the requested content and the content itself both signed with the producer's key. The Data packet follows the path created by the Interest packet on its way back to the consumer.

Neither Data nor Interest contains any kind of information (noticeably, IP addresses) related to the hosts they cross. Interest packets are routed toward data producers based on the name carried in the packet itself. Data packets are then returned based on stated information collected by the Interest packet at each router hop.

NDN routers keep both Interest and Data for some period of time. If an Interest is requested more than once, only the first is sent toward the data source. Routers store Interest packets in Pending Interest Tables (PITs). Each PIT entry contains an identifier for the Interest and a set of interfaces from which the matching Interest has been received. When a Data packet arrives, the router finds the matching PIT entry and forwards it to the interfaces listed in the PIT entry.* In this way the PIT provides an inherent support for multicast.

To achieve optimal usage entries must be timed out quickly, but if entries are timed out prematurely the corresponding incoming Data packets are simply dropped. The consumer takes the responsibility of retransmitting the unsatisfied Interest packet.

A Data packet can be thought of as totally independent from its source and also from its consumer. This way, any router can cache Data packets to satisfy future Interest requests.

Each router is equipped with a Content Store, that is, a cache similar to the buffer memory of today's IP router. Thus, both IP and NDN routers store packets in buffer memory. IP routers cannot, however, reuse data after forwarding them, whereas NDN routers can.† Upon receiving an Interest packet an NDN router first checks its Content Store. If the buffer contains a Data packet whose name falls under the Interest name, the data are sent back as a response. A cache replacement policy ensures the removal of stale Data packets.

* Note that since Data and Interest follow on the same path but in opposite directions, each Data satisfies (i.e., consumes) an Interest at each hop, thus achieving a "hop-by-hop" flow balance.

† Caching named data might raise privacy concerns. Because NDN explicitly names the data, a router owner could find out the name of data as well as that of the content monitoring the router's Content Store. NDN maintains anonymity by removing all information indicating who is requesting the data (maintaining at most only one hop of the path).

Built-in caching enables NDN to natively support the many functions (e.g., content distribution, multicasting, mobility management, delay-tolerant networking) without relying on additional or external infrastructures. Furthermore, by caching content (typically in proximity to the requester), NDN improves packet delivery performance and reduces dependency on a particular source that may fail due to faults or attacks. For static files NDN routers achieve close to optimal data delivery. But even dynamic content can benefit from caching as in the case of multicast communication or packet retransmission after a packet loss.

Naming

In NDN the naming schema is free and left to the application layer. It should still comply with some basic rules, however.

Names are *opaque* to the network. Other than the name format, the Router does not need to know the name scheme to perform its operation correctly. Naming schemes may evolve separately from the network and are a result of conventions established between producers and consumers as seen in today's Representational State Transfer (REST) architecture where naming schemes are typically left to the application. As in REST, names may even refer to a content whose *representation* (using the REST terminology) does not exist at the time the request is issued but is created on the fly once the producer receives the request.

Nevertheless, NDN design guides schema implementers to adopt a hierarchically structured name. In CCNx, for instance, names and components have arbitrary lengths, but component division is explicitly defined. A picture may have the name:

```
ccnx://example.org/pictures/sample.png
```

where the scheme name is conventionally set to ccnx://.

This schema conventionally uses slashes to mark the boundaries between name *components*. A format like this is useful to loosely represent the relationship between content and pieces of the same content (or chunks). For example, version 2 of the previously referenced picture may be assigned the name:

```
ccnx://example.org/pictures/sample.png/2
```

and its chunks might be progressively numbered. The following name is assigned to the fourth chunk of the picture sample.png (version two):

```
ccnx://org/example.org/pictures/sample.png/2/4
```

In this way a consistent hierarchical schema helps consumers request correct interest packets. Relative operators (e.g., next sibling, first child) can also be used as annotations in Interest packets. Since content may be dynamically generated, there must be a way to purposefully generate matching names on the side of both the producer and consumer. Deterministic algorithms can be adopted to allow both producer and consumer to create the same names for content. Alternatively, based on a priori knowledge it is possible to retrieve data with a partial name, similar to what happens in the IP Multimedia Subsystem (IMS) when an Extensible Markup Language (XML) Document Management (XDM) client retrieves data from a server based on partial knowledge of the target Uniform Resource Identifier (URI).

Last but not least, an important feature offered by a hierarchical naming schema is scalability. In fact, even if routing is feasible on flat names (Caesar et al. 2006), hierarchical names enable aggregation. This feature is currently used in IP network addressing and is essential for today's routing table to scale. Hosts are grouped in networks, and outside that network routers store only one single network address (the network prefix) rather than multiple host addresses. Similarly, content can be grouped into collections and assigned names based on a hierarchical naming schema so that routing algorithms can correctly route packets by name prefix, identifying a collection rather than a single content in the collection. This way routing tables save storage space and ensure faster processing.

While routing protocol depends on the hierarchical structure of names, name components may contain arbitrary bytes that generally are not interpreted at the network level.* These bytes may even represent encrypted data, ensuring support for privacy. Conveniently,

* CCNx provides only one exception to this rule. Each name always contains a value that is derived from data as the very last component. This last component, called the digest component, matches the SHA-256 of the entire encoded chunk to which the name refers and is used to identify and suppress potential multiples of packets containing the same chunk. However, since this component can be computed from the data contained in the packet, it is removed before transmission.

noncharacter bytes are converted into characters following the convention defined for URI in RFC 3986 (percent encoding; Berners-Lee, Fielding, and Masinter 2005).

Security

NDN applications benefit from built-in security. Security is built into the datum itself rather than being a function of where or how it is obtained. This paradigm moves the consumer's trust in the datum from its source to the datum itself.

Each piece of datum or chunk of data is signed. That along with its name creates a secure association between the two. The signature coupled with the publisher's information enables the determination of data provenance. Granular security is also supported, allowing the consumer to judge if a data unit is from a trusted publisher or not.

Advantages

NDN does routing and forwarding based on names. This eliminates a number of problems with its IP addressing schema. For example, since the content namespace is not restricted to fixed length addresses, there is no address space exhaustion problem. In IP, this problem has been partially solved by introducing Network Address Translation (NAT)—and then with a slow and difficult transition to Internet Protocol, version 6 (IPv6). In CCNx, there would be no need for any NAT as publishing content is as simple as just giving the content to the network rather than maintaining a publicly exposed IP address at its source.

NDN also allows for seamless mobility, while IP network mobility requires the user's device to change its IP address resulting in a loss of connection. NDN does not use host identifiers; rather, it relies on consistent content names. In this way communication is not broken even if the user's device needs to reestablish a new connection.

NDN packets are signed. They cannot be spoofed or tampered with, and since they do not contain a host address, hosts cannot be addressed. This increases security by making it difficult to send malicious packets to any particular target host.*

* The most effective attack on an NDN network is the Denial of Service, which is extensive and therefore less frequent than others.

Routing and Forwarding

In NDN, routing is accomplished similarly to IP routing, but instead of announcing IP prefixes an NDN router announces name prefixes that cover the data (content or chunk) the router is able to serve. The announcements are disseminated through the network via a routing protocol where each router builds its FIB based on these prefix announcements.

Generally speaking, routers treat names as a sequence of opaque components and simply make the longest component-wise prefix match of them from a packet in the FIB.

To allow a router to distinguish unhandled Interest packets from already routed ones, a random nonce is added to each packet. The router discards duplicate Interest packets to ensure that one copy of each packet at most traverses any link and that no loop is generated. Since Data packets follow the Interest route but in the opposite direction, the Data packets cannot go into a loop. In this way NDN routing is loop free and can send Interest packets on multiple interfaces without worrying about loops. This NDN employs multipath routing as opposed to IP routers, which generally adopt a single best path to avoid loops. For example, an NDN router may forward an Interest packet to all interfaces, measure the performance of returning Data packets, and finally choose the best interface to send the next Interest packets. Or, in the event of a malicious attack, the router can detect anomalies in packets caused by prefix hijacking techniques and try a different path to retrieve the data. These capabilities are called *NDN forwarding strategies* and are responsible for selecting which and how many interfaces have to be used to request an Interest packet as well as how many unsatisfied Interest packets are allowed and in which order they should be satisfied (i.e., their relative priority).

The NDN architecture does not provide a separate transport layer. It is a network designed to operate on top of an unreliable transport service and support the acutely dynamic connectivity of mobile and ubiquitous computing.

Currently, transport protocol functions are placed in the application layer as well as in the forwarding plane strategy. Data integrity and reliability are moved up and are directly handled by application processes, and traffic load is instead managed at the strategy level by regulating PIT size on a hop-by-hop basis. If a node is overloaded by traffic from a neighbour, it can simply slow down or terminate the sending of interest to that interface. This feature has the advantage of eliminating the end host's dependence on the present TCP/IP congestion control. In NDN, congestion may occur only at the node at the end of a congested link, allowing retransmission to take place from there to the consumer unlike in today's IP networks where retransmission must backtrack all the way to the producer.

CCNx

CCNx is an example of an overlay NDN solution that transports content in binary-encoded XML packets of variable length over the current IP technology.

CCNx presents only two types of messages: Interest and Data. An Interest message is a request for content specified by name. The name in the Interest message can be either a specific piece of data (a chunk) or a prefix representing a collection of data requested by the consumer in response. Only the name and a nonce (used by the routing algorithm to avoid loops) are required by an Interest message. Other distinctions can be added to the request to support more advanced features. For instance, some attributes allow a consumer to specify the Interest's lifetime and number of hops, thus limiting in time and space the scope of the Interest. Others enable the consumer to include (or exclude) content based on their relative position, their name, and the publisher's identity.

A Data message is used to supply content. Together with the payload, it carries the content name, the publisher's identification, and a signature. A Data packet is required to contain a valid signature. Other than verifying that the content is actually sent by the publisher, the signature, working as a digest, is used to check data integrity. To verify a signature, all any consumer needs to know is the publisher's public key. In theory, public keys, like other content in CCNx, can be requested without an Interest. While this is indeed possible it uncorks a classic key exchange problem, that is, verifying whether the public key actually belongs to its supposed owner. Therefore, applications should supply a separate trust management infrastructure (e.g., a public key infrastructure or web of trust) for this purpose.

Communication in CCNx is receiver driven. A consumer of data may transmit an Interest message over any available connectivity. Any network node receiving an Interest and having data that satisfy the request may transmit back the Data. The receiver driven transmission requires a one-to-one mapping of Interest and Data so the consumer can, like any other node in the network, regulate the Interest transmission rate and achieve flow balance. Pipelining or transmitting multiple Interest messages without waiting for the corresponding Data messages in response is possible, but it obviously requires the consumer to know in advance ample information about the name structure of the content (or collection of content) to retrieve it.

As CCNx does not rely on a transport protocol, consumers need to set a retransmission timer to ensure that timed-out Interests are properly retransmitted.

CCNx nodes provide buffering/caching and loop-free forwarding of Interest packets. Each node implements a Content Store, an FIB, a PIT, and a Face Manager. A *Face* is a generalization of the interface concept and represents either an actual connection to a network interface that allows the user to send broadcast or multicast packets or a connection to a user application that communicates with the CCNx node. Every message in CCNx passes through at least one Face.

Whenever an Interest message arrives on a node, a lookup is performed in the Content Store. If matching Data is found, it is transmitted through the arrival face as the response to the incoming Interest. With the match found, the Interest is considered satisfied and no further processing is necessary on the node. Any unsatisfied Interest message is looked up in the PIT. If the same Interest entry is already pending, its arrival Face is added to the list of sources for the unsatisfied Interest, and it is dropped. If there is no matching entry in the PIT, the Interest is likely accepted and inserted as a new entry in the PIT (it may also be discarded, depending on the node's load-balancing policy). A lookup is then performed on the FIB to retrieve the destination Faces according to the strategy rules. If the FIB lookup fails, the Interest is dropped from the PIT.

Whenever a Data message arrives on a node, the node first looks up the Content Store. If the Content Store already contains a matching content, the duplicate Data message is simply discarded.

The node then looks up the PIT, and, if a matching Interest is found, the Data are forwarded to any associated arrival Face. If no match is found, the Content can be stored as a new entry in the Content Store to satisfy future Interests.

Security is mainly implemented via content signatures; however, they are a major reason for delays. A signed content contains four concatenated fields: Signature, Name, Signed-Info, and Content.* The Signature field also contains the name of the cryptographic algorithm used to generate the signature, the digital signature itself, and any additional information necessary to verify the signature, in particular an optional Witness field, included when the signature generation is performed over multiple content objects at once (Becker 2008). This enables an individual content to be verified as part of a wider collection.

* Obviously the signature itself is computed over the last three fields.

SECTION IV
Next-Generation Networks

The Internet Protocol (IP) Multimedia Subsystem (IMS) was originally defined in 3rd Generation Partnership Project (3GPP) Release 6 specifications as overlay architecture on top of the 3GPP Packed Switched Core Networks. Its original purpose was to provide real-time multimedia services for wireless networks. Thanks to its widely abstract architecture definition, IMS specifications have been endorsed by other standardization bodies, contributing to the creation of a uniform environment for the development of communication and Internet services.

The European Telecommunications Standards Institute (ETSI) Technical Committee Telecoms and Internet converged Service and Protocol for Advanced Networks (TISPAN)* began to define interfaces to the IMS from fixed-line access technologies, creating the first next-generation network (NGN) converged architecture named TISPAN Release 1.

As new players such as Information Technology (IT) companies and content providers started to enter into the traditional business of device and network suppliers, the NGN architecture progressively became more open. The NGN service layer, which originally included just the IMS, gradually extended to other functional entities implementing access control, charging, user profile management, and other capabilities.

* ETSI Telecoms and Internet converged Service & Protocol for Advanced Networks (TISPAN) Technical Committee, http://www.etsi.org/tispan/.

This section begins with an overview of the IMS and Session Initiation Protocol (SIP). It addresses the role of identities and identifiers in NGNs with special regard for the user identity. The proliferation of identifiers, the need for a uniform authentication mechanism, and the current solution, based on the Generic Authentication Architecture (GAA), are analyzed in details. The exposition concludes with a description of the Extensible Markup Language (XML) Document Management (XDM), the technology that allows uniform management of user- and service-related data in the NGN service layer.

14

Evolution of the Cellular Telephony Networks

Over the last few years the telecoms industry has seen unparalleled growth in fixed-mobile converged services and networks. Fixed-mobile convergence presents the integration of wireline and wireless technologies and services to create a single telecommunications network environment. It promises to overcome some of the physical barriers that now prevent telecom service providers from reaching all of their potential customers with all types of services.

Out of the wireless standards consortium called the 3rd Generation Partnership Project (3GPP) comes a slow-growing and complicated collection of carrier network functions and processes that collectively are referred to as the Internet Protocol (IP) Multimedia Subsystem (IMS). The IMS standards promise an operator-friendly environment for real-time, packet-based calls and services that will not only preserve traditional carrier controls over user signaling and usage-based billing but will also generate new revenue via deep-packet inspection of protocols, the Unified Resource Identifier (URI), and content. The IMS was conceived for the evolution of cellular telephony networks, but the benefits of user signaling and billing controls have attracted the endorsement and involvement of wireline network operators and standards makers, including the European Telecommunications Standards Institute (ETSI), especially its Telecoms and Internet converged Service and Protocol for Advanced Networks (TISPAN) Technical Committee (TC), the U.S.-based Alliance for Telecommunications Industry Solutions (ATIS), and the International Telecommunication Union's Telecommunication Standardization Sector (ITU-T). Other standardization organizations and forums such as the Internet Engineering Task Force (IETF), CableLabs, MultiService Forum (MSF), and Open Mobile Alliance (OMA) are actively involved in defining Next Generation Network (NGN) standards.

NGN Functional Architecture

TISPAN developed a functional architecture (ETSI ES 282 001; ETSI 2005) consisting of a number of subsystems and structured in a service layer and an IP-based transport layer. This subsystem-oriented architecture enables new subsystems to be added over time to cover new demands and service classes. It also provides the ability to import subsystems defined by other standardization bodies. Each subsystem is specified as a set of functional entities and related interfaces. Figure 14.1 shows the overall NGN functional architecture.

The transport layer provides the IP connectivity for NGN users. The transport layer is composed of a transport control sublayer on top of transfer functions. The transfer control sublayer is further divided into the Network Attachment Subsystem (NASS) and the Resource and Admission Control Subsystem (RACS).

The NASS provides registration at the access level and initializes terminal accessing to NGN services (ETSI ES 282 004; ETSI 2010). There may be more than one NASS to support multiple access networks.

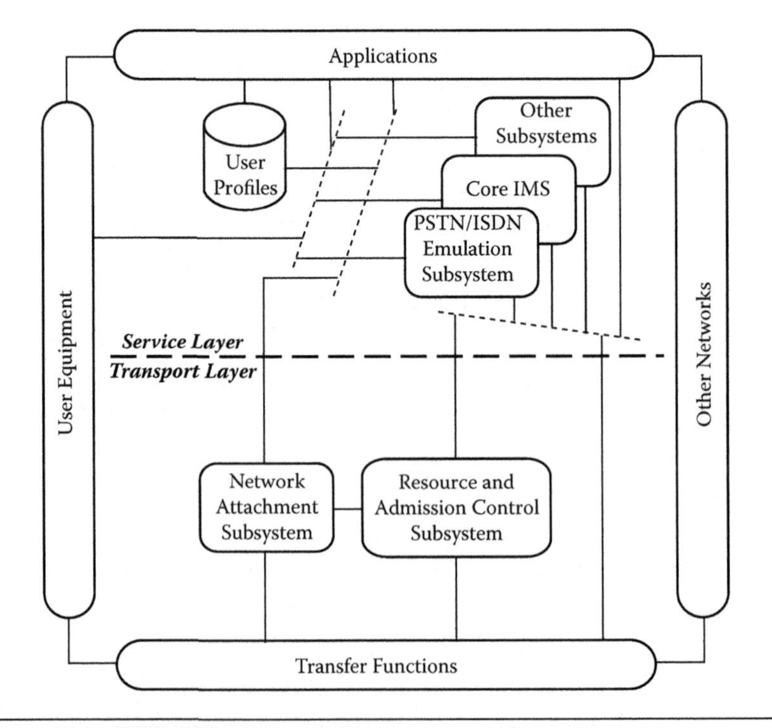

Figure 14.1 TISPAN NGN functional architecture—concept of subsystems.

The RACS provides applications with a mechanism for requesting and reserving resources from the access network (ETSI ES 282 003; ETSI 2011a).

The NGN service layer is composed of the following:

- Core IMS (ETSI ES 282 007; ETSI 2008a)
- Public switched telephone network (PSTN)/Integrated Services Digital Network (ISDN) Emulation Subsystem (PES) (ETSI ES 282 002; ETSI 2006a)
- Other multimedia subsystems (e.g., NGN integrated IPTV Subsystem) (ETSI TS 182 028; ETSI 2008c) and applications
- Common components (i.e., used by several subsystems) such as those required for accessing applications, charging functions, user profile management, security management, and routing databases (e.g., ENUM; RFC 6116; Bradner, Conroy, and Fujiwara 2011)

The functional entities that make up a subsystem may be distributed over network and service provider domains (Figure 14.2). The network attachment subsystem may be distributed between a visited and a home network. Service-layer subsystems that support nomadism may also be distributed between a visited and a home network.

An access network is composed of an access segment and an aggregation segment (Figure 14.3). The access segment (also known as the *last-mile segment*) stretches from the customer premises to the first network node (also known as the *access node*). The aggregation segment is composed of the transport network elements enabling one or more access nodes to be connected to a core network through an IP edge router (Di reference point).

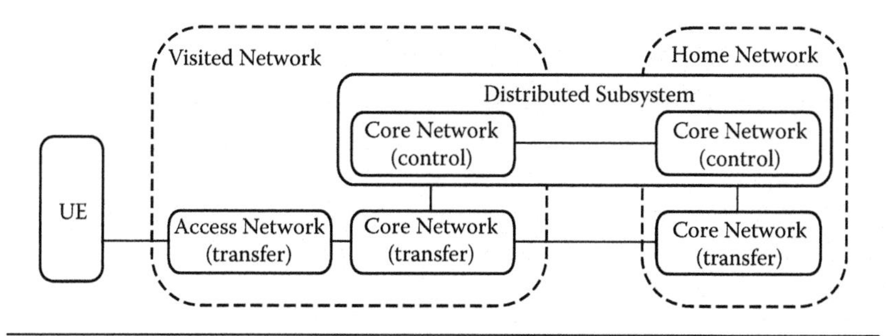

Figure 14.2 Distributed subsystems.

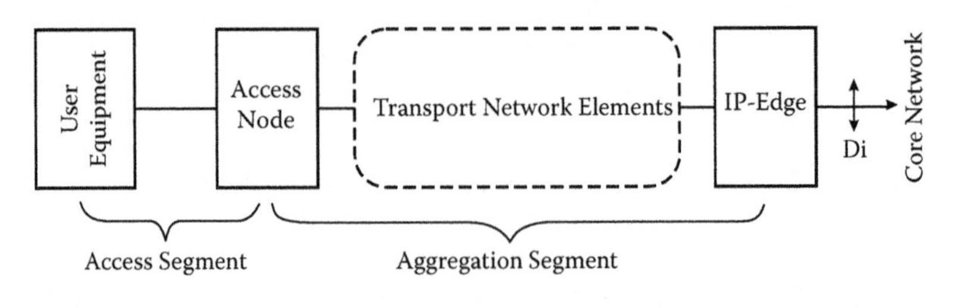

Figure 14.3 Access and aggregation segments.

In configurations where the access segment uses the Digital Subscriber Line (DSL) technology, the aggregation segment generally uses Asynchronous Transfer Mode (ATM) or Gigabit Ethernet. The IP edge is known as a Broadband Remote Access Server (BRAS) or Broadband Network Gateway (BNG), and the access node is a digital subscriber line access multiplexer (DSLAM).

In configurations where the access segment uses the Gigabit-capable passive optical network (GPON) technology, the aggregation segment uses Gigabit Ethernet. The IP edge is known as a BNG, and the functions of the access node are distributed between the Optical Line Termination (OLT) and the Optical Network Unit (ONU) and Optical Network Terminal (ONT).

The core network of NGN is based on the IMS, as defined in 3GPP Release 6 and 3GPP2 revision A for IP-based multimedia applications. The IMS is IP end-to-end and allows applications and services to be supported seamlessly across all networks. IMS is framework architecture—a set of capabilities specified in 3GPP documents that defines components, services, and interfaces for NGN. It uses Voice over IP (VoIP) based on a 3GPP standardized implementation of the Session Initiation Protocol (SIP), and it runs over the standard IP. 3GPP has enhanced and refined the SIP- and IP-based protocols, primarily the Diameter base protocol (RFC 3588; Calhoun et al. 2003), which provides an Authentication, Authorization, and Accounting (AAA) framework for applications such as network access or IP mobility. Diameter is also intended to work in local AAA and roaming situations.

TISPAN has adopted the IMS and had been closely working with 3GPP on any modifications or improvements that may be needed for the NGN.*

* In 2012, the termination of TISPAN was approved, along with transferring the work to the new ETSI Network Technologies (NTECH) TC.

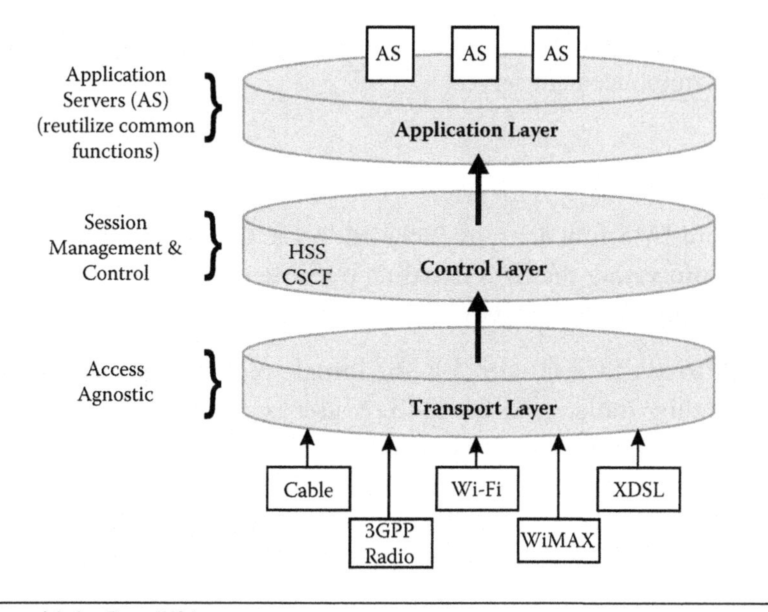

Figure 14.4 Three IMS layers.

There are three distinct operational layers (or planes) within the IMS architecture: the application layer, control layer, and transport layer (Figure 14.4).

The application layer contains a number of application server types. These are all SIP entities as expected within the IMS architecture. These servers host and execute services and can operate in a number of SIP functional modes (RFC 3261; Rosenberg et al. 2002):

- SIP user agent
- SIP back-to-back user agent
- SIP proxy server

The control layer deals with session signaling and contains a number of distinct functions to process the signaling traffic flow, such as the Call Session Control Functions (CSCF), Home Subscriber Server (HSS), Media Gateway Control Function (MGCF), and Media Resource Function Controller (MRFC). Using protocols such as SIP, Diameter, and the Gateway Control Protocol (GCP; ITU-T H.248.1; ITU 2005), the various elements are able to establish subscriber requested services.

The transport layer transports the media streams directly between subscribers and between subscribers and IMS media generating functions, such as the media resource function processor acting as a media announcement server.

IMS Components

The core IMS functions are included in the CSCF, which is an SIP server processing the IMS signaling traffic to control multimedia sessions. There are three types of CSCFs:

1. Proxy CSCF (P-CSCF): The initial point of contact for signaling traffic in to the IMS. A user is allocated P-CSCF as a part of the registration process and provides a two-way IPsec association with the user; all signaling traffic traverses P-CSCF for the duration of the session.
2. Serving CSCF (S-CSCF): Provides the service coordination logic to invoke and orchestrate the application servers needed to deliver the requested service. S-CSCF interacts with the HSS to determine user service eligibility by downloading the user profile; the S-CSCF is allocated for the duration of the registration.
3. Interrogating CSCF (I-CSCF): An SIP proxy that provides a gateway to other domains, such as other service provider networks. I-CSCF may encrypt sensitive domain information, a function referred to as a Topology Hiding Internetwork Gateway (THIG), before forwarding the traffic.

Until Release 6, specifications for the P-CSCF included the Policy Decision Function (PDF), which stores policies and consults them to make decisions about IP bearer resource allocation requests. The PDF has been separated from the P-CSCF to make it more accessible to Wireless Local Area Networks (WLANs) and other access network types. P-CSCFs also generate Call Detail Records (CDRs) or billing records that can be consolidated at a Charging Gateway Function (CGF).

The 3GPP IMS has been extended in the TISPAN NGN to support additional access network types, such as xDSL and WLAN.

The TISPAN extensions of the 3GPP IMS (Figure 14.5) take into account the differences between the wireless and wired environments, especially as far as the amount of control needed for the end-user

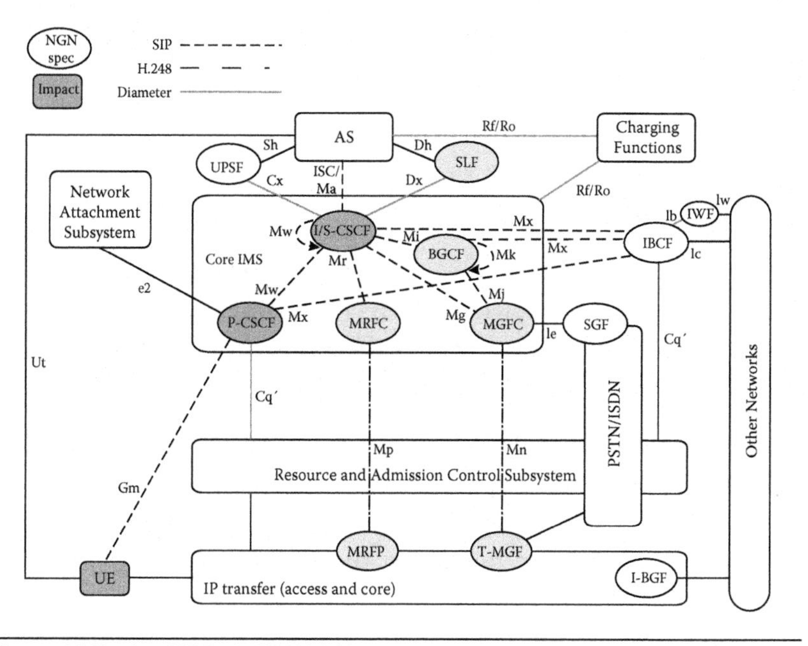

Figure 14.5 Core IMS in the TISPAN NGN.

device that leads to huge differences in perspectives on control and management is concerned. As a more specific example, there is no equivalent in wireline networks to the so-called smart cards used to identify mobile customers and trigger billing activity when they make calls from another wireless operator's service territory. For both environments, there are also different regulatory constraints, quality of service and location definitions, methods and mechanisms, and support requirements for legacy devices, as well as many differences in security and network management details.

Other NGN Subsystems

PES supports the emulation of PSTN/ISDN services for legacy terminals connected to the NGN through residential gateways or access gateways. Further details about the functionalities and architecture of PES can be found in ES 282 002 (ETSI 2006a), which defines alternative functional architectures for this subsystem.

The IPTV subsystem supports the provision of content on demand services and broadcast services using a dedicated service control architecture described in TS 182 028 (ETSI 2008c). The TISPAN architecture also enables support for IPTV services using the IMS, as described in TS 182 027 (ETSI 2011b).

Common Components

Common components are those that can be accessed by more than one subsystem. Two types of common components can be identified: (1) components known in 3GPP IMS and (2) new components defined by TISPAN. The first group of components has been defined by 3GPP IMS and includes the following functions:

1. User Profile Server Function (UPSF), which is, in fact, a subset of the HSS defined by 3GPP. The HSS is the master database for a given user. It is the entity containing the subscription-related information to support the network entities actually handling calls or sessions. As an example, the HSS provides support to the call control servers to complete the routing and roaming procedures by solving authentication, authorization, naming, and addressing resolution and location dependencies. UPSF in NGN stores all relevant information regarding the user, including identification, addressing, numbering, access controls, and location information. UPSF is similar to HSS in many aspects. The main difference is that HSS is an evolution of the Global System for Mobile Communications (GSM) home location register (HLR), which includes the HLR authentication center (AuC) functionality that provides mobility management. This is not required in NGN; hence, the UPSF is limited to IMS-specific parts of HSS.
2. Subscription Location Function (SLF) is needed only when multiple HSSs are used. Within TISPAN, it can be accessed by service control subsystems and Application Server Functions (ASFs) to retrieve the identity of the UPSF containing the service-level user profile of a particular subscriber.
3. ASF offers value-added services and resides either in the user's home network or in a third-party location. The third party could be a network or simply a stand-alone application server.

4. Interconnection Border Control Function (IBCF) provides the interconnection with other multimedia subsystems.
5. Charging and Data Collection Functions: As the name suggest, these provide data collection and billing mediation for online and offline charging.

The second group represents either new components that have been defined by TISPAN or those 3GPP ones that have been modified by TISPAN in the context of NGN:

1. ASF may provide stand-alone services or value-added services on top of a basic session. For resource control purposes in NGN, ASF Type 1, the first category, may interact with the RACS, while the second category, ASF Type 2, relies on the control subsystem that provides the basic session over which the valued-added service is built. Examples of ASFs are SIP application servers and open services architecture (OSA) application servers. When sitting on top of the IMS, the second type of ASF is identical to the application server function defined by 3GPP, although a network node implementing this functional entity in an NGN network and a network node implementing it in a 3GPP network may differ in terms of supported services.
2. Interworking Function (IWF) is a new component that performs the interworking between protocols used within TISPAN NGN service control subsystems and other IP-based protocols (e.g., between the SIP profile used in the IMS and other SIP profiles or IP-based protocols such as the H.323 protocols; ITU-T H.323; ITU 2009).
3. Charging and Data Collection Functions include data collection functions and mediation functions to the billing systems (for supporting both online and offline charging) or other management applications that may use the same data. Charging in TISPAN Release 1 is limited to offline charging only. There are several functional entities in the Core IMS that may act as charging trigger points: application server; Border Gateway Control Function (BGCF);

(I-/P-/S-) CSCF; MGCF; and MRFC. Moreover, TISPAN has defined the IBCF to which the Core IMS is connected as another kind of charging trigger point. The specification of a subsystem-independent generic architecture of the charging and data collection functions is outside the scope of the current TISPAN release.

15

SESSION INITIATION PROTOCOL

The Session Initiation Protocol (SIP) emerged in the mid-1990s from the research of Henning Schulzrinne, associate professor in the Department of Computer Science at Columbia University, and his team. A coauthor of the Real-Time Transport Protocol (RTP) for transmitting real-time data via the Internet, Schulzrinne also co-wrote the Real-Time Streaming Protocol (RTSP), a proposed standard for controlling streaming audiovisual content over the web. Schulzrinne's intent was to define a standard for Multiparty Multimedia Session Control (MMUSIC). In 1996, he submitted a draft to the Internet Engineering Task Force (IETF) that contained the key elements of SIP. In 1999, he removed extraneous components regarding media content in a new submission, and the IETF issued the first SIP specifications, RFC 2543 (Handley et al. 1999). While some vendors expressed concerned that protocols such as H.323 (ITU 2009) and Gateway Control Protocol (GCP) could jeopardize their investments in SIP services, the IETF continued its work and issued SIP specifications (RFC 3261; Rosenberg et al. 2002) in 2001. The advent of RFC 3261 signaled that the fundamentals of SIP were in place. Since then, enhancements to security and authentication, among other areas, have been issued in several additional RFCs. RFC 3262 (Rosenberg and Schulzrinne 2002b), for example, governs the reliability of provisional responses. RFC 3263 (Rosenberg, and Schulzrinne 2002c) establishes rules to locate SIP proxy servers. RFC 3264 (Rosenberg and Schulzrinne 2002a) provides an offer–answer model, and RFC 3265 (Roach 2002) determines specific event notification.

As early as 2001, vendors began to launch SIP-based services. Since that time, the enthusiasm for the protocol has been growing. Organizations such as Sun Microsystems' Java Community Process are defining application programming interfaces (APIs) using the popular Java programming language so developers can build SIP components and applications for service providers and enterprises. Most importantly, increasing numbers of players are entering the SIP marketplace with promising new services, and SIP is on track to become one of the most significant protocols since the Hypertext Transfer Protocol (HTTP) and the Simple Mail Transfer Protocol (SMTP). SIP technology is still evolving, but the core specifications have been fully standardized, adopted, and widely implemented. Many commercial products and communication platforms on the market are SIP based, including those from Cisco™, Microsoft®, Avaya®, and Radvision®. SIP also brings the Internet approach to multimedia real-time communication; thus, SIP communication may be seen as just another Internet Protocol (IP) service, running on standard hardware and standard operating systems and middleware. Thanks to this approach, many high-performance and high-quality software products are available under some form of open-source or free software licenses. These products together with the appropriate software packages can be used to create a rich multimedia converged network (Segec and Kováčiková 2011).

SIP Features

SIP is an application layer control protocol for initiating, managing, and terminating multimedia sessions across IP networks. SIP sessions involve one or more participants and can use unicast or multicast communication. Borrowing from ubiquitous Internet protocols such as HTTP and SMTP, SIP is text encoded and highly extensible. It may be extended to accommodate features and services such as call control services, mobility, and interoperability with existing telephony systems.

Like the Internet, SIP is easy to understand, extend, and implement. As an IETF specification, SIP extends the open-standards spirit of the Internet to messaging and enabling disparate computers, phones, televisions, and software to communicate. As noted, an SIP message is very similar to HTTP (RFC 2068; Fielding et al. 1997). Much of the syntax in message headers and many HTTP codes are reused. Using SIP, for example, the error code for an address not found, 404, is identical to the web's. SIP also reuses the SMTP for address schemes. An SIP address, such as sip: john@abinstitute.com, has the exact structure of an email address. SIP even leverages web architectures, such as Domain Name System (DNS), making messaging among SIP users even more extensible. Using SIP, service providers can freely choose among standards-based components and quickly harness new technologies. Users can locate and contact one another regardless of media content and numbers of participants. SIP negotiates sessions so that all participants can agree on and modify session features. It can even add, drop, or transfer users.

However, SIP is not a cure-all. It is not a session description protocol, nor does it provide conference control. When initiating multimedia teleconferences, Voice over IP (VoIP) calls, streaming video, or other sessions, there is a requirement to convey media details and to transport addresses and other session description metadata to the participants. The Session Description Protocol (SDP) provides a standard representation for such information, irrespective of how that information is transported. SDP is purely a format for session description. This means that it does not incorporate a transport protocol; it is rather intended to use different transport protocols as appropriate, for example, SIP.

SIP also does not itself provide Quality of Service (QoS) and interoperates with the Resource Reservation Protocol (RSVP) for voice quality. It also works with a number of other protocols, including the Lightweight Directory Access Protocol (LDAP) for location, the Remote Authentication Dial In User Service (RADIUS) for authentication, and RTP for real-time transmissions, among many others.

In general, SIP supports five functions for establishing and terminating multimedia communications:

1. User location: identification of the end system to be used for communication
2. User availability: identification of the willingness of the called party to engage in communications
3. User capabilities: identification of the media and media parameters to be used
4. Session setup: "ringing," establishment of session parameters at both called and calling party
5. Session management: functions such as transfer and termination of sessions, modifying session parameters, and invoking services

An important feature of SIP is that it does not define the type of session that is being established, only how it should be managed. This flexibility means that SIP can be used for a high-number number of applications and services, including interactive gaming, music, and Video on Demand (VoD) as well as voice, video, and web conferencing.

SIP Entities

SIP sessions utilize up to four major components (Pavel 2010). Each entity has specific functions and participates in SIP communication as a client (initiates requests), as a server (responds to requests), or as both. One *physical device* can have the functionality of more than one logical SIP entity. For example, a network server working as a proxy server can also function as a registrar at the same time.

Following are the four types of logical SIP entities:

1. SIP User Agents (UAs) are the end-user devices, such as cell phones, multimedia handsets, personal PCs, and PDAs, used to create and manage an SIP session. The User Agent Client (UAC) initiates the message. The User Agent Server (UAS) responds to it.
2. SIP registrar servers are databases that contain the location of all UAs within a domain. In SIP messaging, these servers retrieve and send participants' IP addresses and other pertinent information to the SIP proxy server.
3. SIP proxy servers accept session requests made by an SIP UA and query the SIP registrar server to obtain the recipient UA's addressing information. They then forward the session invitation directly to the recipient UA if it is located in the same domain or to a proxy server if the UA resides in another domain.
4. SIP redirect servers allow SIP proxy servers to direct SIP session invitations to external domains. SIP redirect servers may reside in the same hardware as SIP registrar severs and SIP proxy servers.

The experience gained from over ten years of SIP evolution has led to the introduction of new types of SIP entities, which are not a part of the main SIP specification.

One of these new types of SIP entities is a special concept of an SIP proxy-like server, called a Back-to-Back User Agent (B2BUA). The main idea behind the B2BUA development was to have an SIP proxy-like server that is inserted actively into an SIP call. The B2BUA is inserted into SIP dialogue and splits a call (signaling and media) into two legs. It presents itself as a callee to a caller and as the caller to the callee. This allows B2BUA to play a more active role in an SIP network, for example, to perform a number of functions that are not possible by using the SIP proxy only. This includes accurate call accounting, pre-paid rating and billing, fail over call routing, and topology hiding. The B2BUA usually does not perform any media relaying or processing.

A Session Border Controller (SBC) is another kind of newer SIP entity. It is usually placed as a security device at the border of a provider network to exert control over signaling and media streams that are entering into or leaving a provider network. An SBC provides functions such as security enforcement, QoS guarantees, special connectivity handling functions, and media processing.

Media servers are special-purpose network servers that are used for processing media streams associated with an SIP session. Some of the usual deployment examples include media transcoders, voicemail recording machines, and conferencing machines. Due to high processing load when handling media, the hardware of media servers is usually very powerful.

SIP offers many flexible forms for development of new services. For location of an SIP communication service, any of the previously mentioned SIP server entities can usually be used. In addition, a special dedicated server for hosting SIP applications and services has been developed called an SIP application server. The application server is a special-purpose platform that allows for quick development and runtime environments supported by some kinds of Application Programming Interfaces (APIs) dedicated to service development, service hosting, and operation.

The main SIP specification doesn't mention a location server (LS) as a separate entity; rather, it defines only location service, which is used to store user location information. A location server is the implementation of a location service, which provides the information about callers' possible locations to redirect and proxy servers. It may also contain other kinds of records such as call details, credentials for authentication and authorization, and routing information. In general, the location server is a kind of a database (e.g., Standard Query Language [SQL], Lightweight Directory Access Protocol [LDAP]).

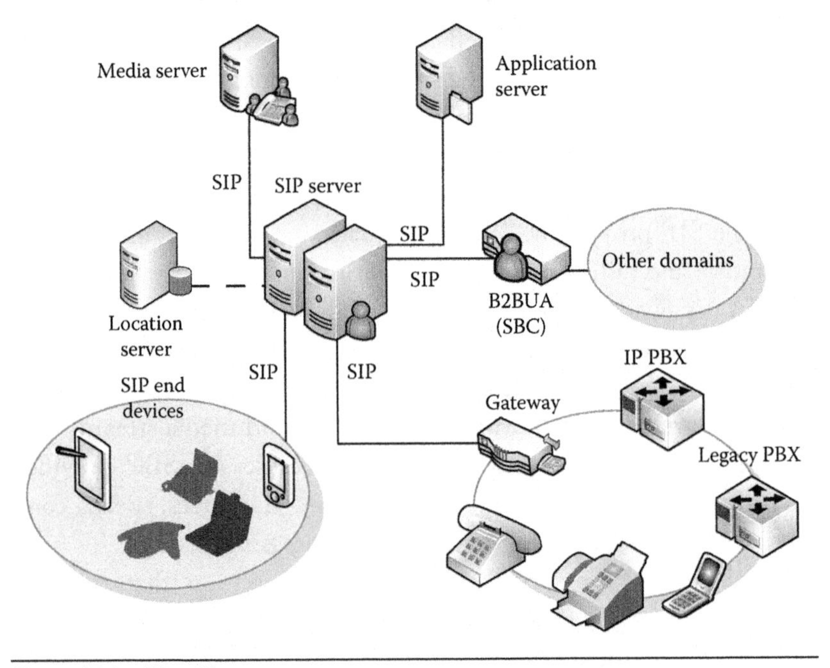

Figure 15.1 SIP entities.

SIP end devices or endpoints are the implementations of an SIP UA, which allow SIP sessions (audio, video, presence, instant messaging) to be initiated and managed. They are implemented as a software-based application, installed and running on some kind of device with IP connectivity (e.g., PC, PDA, 3G phone), or implemented as a dedicated hardware controlled by an SIP UA application (e.g., hardware video telephones, conference communicator).

An overview of SIP logical entities is depicted in Figure 15.1.

Due to the need to interconnect users of SIP-based networks with users of other types of networks working with other protocol sets, a special device sitting at the border of different networks is required. This device is called a gateway. There are many kinds of gateways depending on their functionality or the deployment architecture. A media gateway is a translation device that converts digital media streams between disparate networks. A signaling gateway is a device responsible for converting signaling messages (e.g., information related to call establishment, billing, location, short messages, address conversion) between networks using different signaling protocols (e.g., SIP and H.323, SIP and SS7) as depicted in Figure 15.2. It can also be an integrated device converting both signaling and media. The term SIP gateway is used for

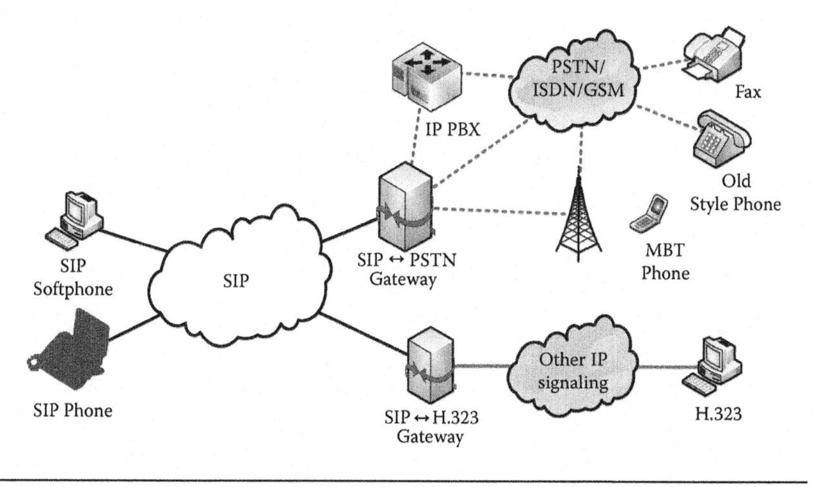

Figure 15.2 SIP-H.323 ad SIP-SS7 gateways.

a special gateway that uses the SIP protocol at the IP side. The main SIP specification understands an SIP gateway as a special-purpose UA that is able to handle many transactions simultaneously.

SIP Messages

Communication using SIP (often called signaling) is composed of a series of messages that can be transported independently by the network. Usually they are each transported in a separate User Datagram Protocol (UDP). Each message consists of a first line, message header, and message body. The first line identifies the type of message. There are two types of messages: request messages and response messages. Requests are usually used to initiate some action or inform a recipient of the request for something. Replies are used to confirm that a request was received and processed and contain the status of the processing.

An example of a typical SIP request message is

```
INVITE sip:userB@serverY.com SIP/2.0
Via: SIP/2.0/TCP PC44.serverX.com:5060;branch=z9hG4bK74af9
Max-Forwards: 70
To: userB <sip:userB@serverY.com>
From: userA <sip:userA@serverX.com>;tag=1928301774
Call-ID: a84b4c76e66710@PC44.serverX.com
CSeq: 325168 INVITE
Contact: <sip:userA@PC44.serverX.com>
Content-Type: application/sdp
Content-Length: 132
```

The first line refers to the method INVITE, which is used to establish a session. The Uniform Resource Identifier (URI) on the first line—`sip:userB@serverY.com`—is called the Request-URI and contains the URI of the next hop of the message (Via, in the second line). In this case it will be the host PC44.serverY.com on port 5060. A branch parameter here identifies this transaction.

Max-Forwards serves to limit the number of hops a request can make on the way to its destination. It consists of an integer that is decremented by one at each hop.

The From and To header fields identify initiator (caller) and recipient (callee) of the invitation, just like in the SMTP where they identify sender and recipient of a message. The From header field contains a tag parameter that serves as a dialog identifier.

The Call-ID header field is a dialogue identifier, and its purpose is to identify messages belonging to the same call. Such messages have the same Call-ID identifier.

CSeq is used to maintain the order of requests. Because requests can be sent over an unreliable transport that can reorder messages, a sequence number must be present in the messages so that the recipient can identify retransmissions and out-of-order requests.

The Contact header field contains the IP address and port on which the sender is awaiting further requests sent by the callee.

Content-Type contains a description of the message body (not shown). Content-Length contains an octet (byte) count of the message body.

The header may contain other header fields also. However, those fields are optional. Please note that the body of the message is not shown here. The body is used to convey information about the media session written in SDP (RFC 4566; Handley, Jacobson, and Perkins, 2006).

There are several request methods in SIP, and new methods are being created all the time by the Internet Engineering Task Force (IETF). The IETF RFC 3261 (basic SIP; Rosenberg et al. 2002) defines the following:

INVITE: It indicates that a user or service is being invited to participate in a call session.

ACK (Acknowledge): It confirms that the client has received a final response to an INVITE request.

BYE: It terminates a call and can be sent by either the caller or the callee.

CANCEL: It cancels any pending searches but does not terminate a call that has already been accepted.
OPTIONS: It queries the capabilities of servers.
REGISTER: It registers the address listed in the To header field with an SIP server. Gateways do not support the REGISTER method.

There are other standard request methods besides the ones mentioned in the basic IETF RFC 3261 (Rosenberg et al. 2002) specification:

PRACK request (RFC 3262; Rosenberg, Schulzrinne 2002b) plays the same role as ACK, but for provisional responses. There is an important difference, however. PRACK is a normal SIP message, like BYE. As such, its own reliability is ensured hop by hop through each stateful proxy. Also like BYE, but, unlike ACK, PRACK has its own response.
REFER (RFC 3515; Sparks 2003) indicates that the recipient (identified by the Request-URI) should contact a third party using the contact information provided in the request. Unless stated otherwise, the protocols for emitting and responding to a REFER request are identical to those for a BYE request in the basic SIP specification. A REFER request implicitly establishes a subscription to the referred event.
PUBLISH (RFC 3265; Roach 2002) is similar to REGISTER in that it allows a user to create, modify, and remove state in another entity that manages this state on behalf of the user. Addressing a PUBLISH request is identical to addressing a SUBSCRIBE request.
SUBSCRIBE and NOTIFY (RFC 3265; Roach 2002) enable entities in the network to subscribe (SUBSCRIBE) to resource or call state for various resources or calls in the network, and those entities (or entities acting on their behalf) can send notifications (NOTIFY) when those states change.
MESSAGE (RFC 3428; Campbell et al. 2002) allows the transfer of instant messages. MESSAGE requests carry the content in the form of MIME body parts. MESSAGE requests do not themselves initiate an SIP session; under normal usage each instant

message stands alone, much like pager messages. MESSAGE requests may be sent in the context of a session initiated by some other SIP request.

When a user agent or proxy server receives a request, it sends a reply. Each request must be replied except ACK requests, which trigger no replies. A typical SIP response looks like the following:

```
SIP/2.0 200 OK
Via: SIP/2.0/UDP PC44.serverX.com;branch= z9hG4bK74af9;received=193.0.2.1
To: userB <sip:userB@serverY.com>;tag=a6c85cf
From: userA <sip:userA@serverX.com>;tag=1938301774
Call-ID: a84b4c76e66710@PC.server44.com
CSeq: 325168 INVITE
Contact: <sip:userB@193.0.2.2>
Content-Type: application/sdp
Content-Length: 122
```

Responses are similar to the requests, except for the first line. The first line of response contains protocol version (SIP/2.0), reply code, and reason phrase. The reply code is an integer number from 100 to 699 which indicates the type of the response. There are six classes of responses:

1xx: Provisional—request received, continuing to process the request
2xx: Success—the action was successfully received, understood, and accepted
3xx: Redirection—further action needs to be taken to complete the request
4xx: Client Error—the request contains bad syntax or cannot be fulfilled at this server
5xx: Server Error—the server failed to fulfill an apparently valid request
6xx: Global Failure—the request cannot be fulfilled at any server

In addition to the response class, the first line also contains a reason phrase. The reason phrase usually contains a human-readable message describing the result of the processing.

The request to which a particular response belongs is identified using the CSeq header field. In addition to the sequence number, this header field also contains the method of the corresponding request. In the previous example it was the INVITE request.

SIP Transactions

A transaction consists of one request and all responses to that request. SIP messages are usually arranged into transactions by user agents and certain types of proxy servers (ITU-T H.323; ITU 2009); therefore, SIP is considered to be a *transactional protocol*. The responses will include zero or more provisional responses and one or more final responses (an INVITE might be answered by more than one final response when a proxy server forks the request). If a transaction was initiated by an INVITE request, then the same transaction also includes ACK, but only if the final response was not a 2xx response. If the final response was a 2xx response, then the ACK is not considered a part of the transaction.

As ACK is part of transactions with a negative final response but is not part of transactions with positive final responses, this is asymmetric behavior. The reason for this separation is the importance of delivery of all 200 OK messages, which not only can establish a session but can also be generated by multiple entities when a proxy server forks the request and all of them must be delivered to the calling user agent. Therefore, user agents take responsibility in this case and retransmit 200 OK responses until they receive an ACK. Only responses to INVITE are retransmitted.

SIP entities that have notion of transactions are called stateful. Such entities usually create a state associated with a transaction that is kept in the memory for the duration of the transaction. When a request or response comes, a stateful entity tries to associate the request (or response) to existing transactions. To be able to do so, it must extract a unique transaction identifier from the message and compare it with identifiers of all existing transactions. If such a transaction exists, then its state gets updated from the message.

In RFC 3261 (Rosenberg et al. 2002), an SIP message includes the identifier directly. This is significant simplification in comparison with the previous RFC 2543 (Handley et al. 1999), according to which the transaction identifier was calculated as hash of all important message header fields (that included To, From, Request-URI, and CSeq). This was quite slow and complex. Moreover, during interoperability tests, such transaction identifiers used to be a common source of problems. In RFC 3261, the way of calculating transaction identifiers was changed.

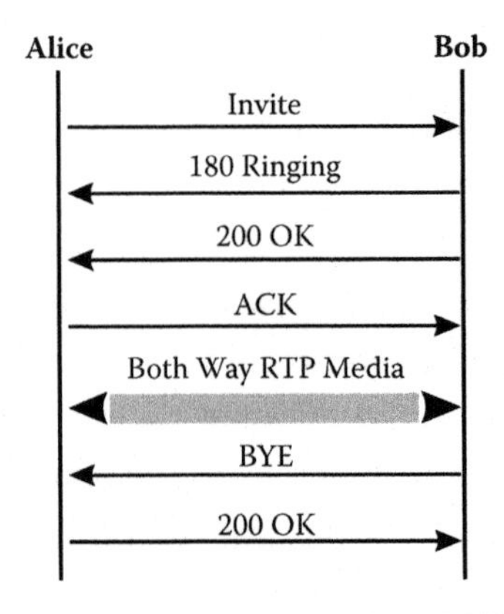

Figure 15.3 Successful session establishment and termination.

Instead of complicated hashing of important header fields, an SIP message now includes the identifier directly. The branch parameter of Via header fields contains the transaction identifier directly.

Example of an SIP Typical Scenario

In the scenario in Figure 15.3, UserA completes a call to UserB directly. A session invitation consists of one INVITE request, which is usually sent to a proxy. The 180 Ringing response is generated when the callee's phone starts ringing. A 200 OK is generated once the callee picks up the phone, and it is retransmitted by the callee's user agent until it receives an ACK from the caller. The session is established at this point. Session termination is accomplished by sending a BYE request within the dialogue established by INVITE. BYE messages are sent directly from one user agent to the other in this example. A party wishing to tear down a session sends a BYE request to the other party involved in the session. The other party sends a 200 OK response to confirm the BYE, and then the session is terminated.

16

Identifiers in Communication Networks

Identity in the wider use of the term does not mean the same as the concept of identity used in telecommunications, that is, the collection of identifiers, permissions, and authentication data necessary to obtain access to services (ETSI TR 187 010; ETSI 2008b). Most current telecommunications identification schemes use a single identifier to perform (at least) two distinct functions:

1. Routing: Identifiers can be processed by information and communication systems to enable end-to-end service instances between endpoints to be established.
2. Identification: End users can identify the source of an incoming communication, such as Calling Line Identification Presentation (CLIP) or email addresses, or can confirm the identity of the remote endpoint to which a connection has or will be established, such as Connected Line Identification Presentation (COLP) and Uniform Resource Locator (URL).

Failure of the first of these two functions may result in loss of service to end users. To ensure that such failures do not occur, rules relating to the content and formatting of communication identifiers are enforced. As a result, most communications-related identifiers have a defined structure that simplifies the identification of region, domain, or endpoint.

In many cases of attack on identity, countermeasures already exist, using corroborating data to reinforce the observation of an assertion of identity. Many organizations do not rely on a single identifier as an assertion of identity. Consequently, when trying to masquerade as a legitimate user, a criminal will seek to recover multiple correlated forms of identification and use them in combination to counter the identity checks. In contexts where identity is represented by an identifier having a known structure (as is the case in email names and the E.164 [ITU 2010] numbering schemes; RFC 6116 [Bradner et al. 2011]), it is possible for the identity to be falsely claimed.

Identity management is an important aspect in overcoming current concerns regarding the definition of exactly what constitutes a user and what rights that user has. Unfortunately, as identity is a rather abstract concept, its management is difficult to define and specify.

Background to Identifiers for Next-Generation Networks

As a design goal, next-generation networks (NGNs) must be able to support the existing naming, numbering, and addressing plans for both fixed and mobile networks (ETSI TS 184 002; ETSI 2006b). This is quite ambitious, considering the variety of networks and their naming–numbering schemas that NGNs aim to integrate. For networks such as the public switched telephone network (PSTN) or the Integrated Services Digital Network (ISDN), for the Global System for Mobile Communications (GSM)–based public land mobile networks (PLMNs), and for the Internet, there is a common terminology used concerning the present identifiers (IDs) used in these networks. The terms, which are defined in ITU-T Recommendation E.191 (ITU-T E.191; ITU 2000a), are name, number, and address. In the PSTN the ID is the E.164 number (ITU-T E.164; ITU 2010), and that number is used for identifying and routing the call to the subscriber–user or service. With the introduction of services based on nongeographic numbers[*] and number portability,[†] the function of the number has been split between a name role for identifying the user or service and an address role to indicate how to route the call to the subscriber's network termination point.

In GSM-based PLMNs, the E.164 (ITU 2010) number is often called a Mobile Station ISDN Number (MSISDN) to indicate that the E.164 number is used for mobile services. Another ID used in GSM networks is the International Mobile Subscriber Identity (IMSI),

[*] Nongeographic numbers are telephone numbers that provide callers with a contact number giving no indication as to the geographical location of the line being called.

[†] Number portability enables a user as long as she remains in the same geographic area to switch telephone service providers, including interconnected Voice over Internet Protocol (VoIP) providers, and to keep her existing phone number.

based on ITU-T Recommendation E.212 (ITU-T E.212; ITU 2008). The IMSI provides a unique identifier of the mobile subscription for registration purposes. It is also used to identify the Home Public Land Mobile Network (HPLMN) when the mobile subscriber–terminal is roaming in a Visited Public Land Mobile Network (VPLMN). Most of the present Subscriber Identity Module (SIM) cards used in GSM networks are marked with another ID called the Issuer Identification Number (IIN) according to ITU-T Recommendation E.118 (ITU-T E.118; ITU 2006).

For circuit-switched networks, there are also some IDs used for different network functions. For example, Signaling Point Codes (SPCs) used in the ITU-T Signaling System No. 7 (SS7). International Signaling Point Codes (ISPCs) are used in international networks according to ITU-T Recommendation Q.708 (ITU-T Q.708; ITU 1999), and National Signaling Point Codes (NSPCs) are used between national networks. These addresses can be seen as public IDs, but some SPCs, with a Network Indicator (NI), are used solely as a private ID inside one operator's network and are therefore never disclosed to other operators.*

For packet-switched networks such as the Internet and other IP-based networks, names in the form of domain names (RFC 1035; Mockapetris 1987b) are used. The separation between a name and an address has been used from the earliest days of the Internet (although originally the IP address was the only major ID). The domain name is used to identify the user–host, and the IP address is used for routing to the interface to which the host is connected (for IP address format IPv4 and IPv6). The IP address is received through a name resolution with the help of the Domain Name System (DNS). Before the growth and success of the Internet, other Public Data Networks (PDNs) based on X.25 (ITU-T X.25; ITU 1996) and X.21 (ITU-T X.21; ITU 1992) were used, and some are still in use. For these PDNs, a numbering plan based on ITU-T Recommendation X.121 (ITU-T X.121; ITU 2000b) is used.

Specific address resources named Asynchronous Transfer Mode (ATM) End System Addresses (AESA) are used in ATM

* The Network Indicator (NI) indicates whether the message is for a national or international network.

networks. Different AESAs are used, and one (ITU-IND AESA) is administered by the International Telecommunications Union Telecommunication Standardization Bureau (ITU TSB) and is based on ITU-T Recommendation E.191 (ITU-T E.191; ITU 2000a).

There are also IDs from other naming, numbering, addressing, or identification plans, such as IDs for the TErrestrial Trunked RAdio (TETRA; ETSI EN 300 392-2; ETSI 2007) that were specifically designed to meet the needs of Public Mobile Radio (PMR), Walkie-Talkie, and more network-specific IDs like Network Service Access Points (NSAPs).

3GPP Concept on the Use of Identifiers

In Universal Mobile Telecommunication Systems (UMTS)–based PLMNs many specific IDs are used, and they are described in detail in the ETSI TS 123 003 (ETSI 2000). The general requirements addressed by these IDs are that the subscriber can be uniquely identifiable by the network, the service network can authenticate the subscriber, and if a subscriber is being served by a network other than his home network, the visiting network shall be able to identify the associated home network. The Universal Integrated Circuit Card (UICC) and the two applications that may reside on it, the universal Subscriber Identity Module (USIM) and the Internet Protocol (IP) Multimedia Services Identity Module (ISIM), are described in more detail following.

With the introduction of the IP Multimedia Subsystem (IMS) domain within the 3rd Generation Partnership Project (3GPP), a single level of registration was found to be insufficient. To provide the necessary functionality for registration to IMS, a new application, the ISIM, was added to the UICC. In summary, the USIM application is used to gain access to the UMTS access network—the Circuit Switched/General Packet Radio Service (CS/GPRS) network—and the ISIM is used to gain access to the IMS domain (3GPP Release 6 or higher).

A subscriber may access the IMS domain (1) with the values of the ISIM identifiers derived from the USIM (in the absence of ISIM); and (2) with the value of the ISIM identifiers provisioned independently from the USIM. Thus, an UICC may contain an ISIM, but this is not mandatory.

The SIM card introduced not only a method of uniquely identifying a subscription irrespective of the GSM device being used but also a higher level of security that was absent from the previous analogue system, which suffered from the relatively easiness of frauds based on handset cloning.

Whereas the SIM is commonly seen as the physical card together with the software to authenticate, authorize, and identify the subscriber, the UICC merely defines the physical characteristics of the smart card. The USIM and ISIM are software applications resident on the UICC. Legally, the UICC remains a property of the mobile network operator who has the authority to decide which applications can reside on it, such as USIM and ISIM. Each operator can use the card as a network endpoint independently from the mobile phone. The UICC is the port of call within each mobile device where mobile operators can store the applications that bring their services to fruition, such as for roaming, branding, device tracking, and browsers. Storing such applications in the UICC means that they benefit from the UICCs' security credentials, thus offering revenue protection for the operator. Being removable, it is easy to transfer them when devices are renewed without the need to rewrite the application.

The UICC is the only operator-owned item that resides in the hands of the end user. It connects into the operator system via Over-The-Air (OTA) technology that enables secure remote management. To take roaming as an example, the combination of UICC and OTA technology allows carriers to update their roaming agreements, remotely, and with no visible impact to the end user.

Examples of the types of applications that can be stored in the UICC include the following:

- High-level applications based on standard APIs, such as Java Card applets
- Applications based on service engines, using standard APIs
- Network access applications: USIM, ISIM
- Applications that require chips with optional features, such as Near-Field Communication (NFC) applications

The ISIM provides access to the IMS via any IP access network. The ISIM enables each subscription to have multiple public identities. It enhances interoperability by reducing the options for implementation as there is no need to accommodate legacy USIM or CDMA Subscriber Identity Module (CSIM) applications. In addition, the ISIM facilitates the provisioning of important parameters across all mobile terminals. It works by providing a set of IMS security data and functions for IMS access: mutual authentication and key agreement, provisioning, and Generic Bootstrapping Architecture (GBA) for IMS-based services.

Identity and Identifiers in NGN

An identity is used within the NGN to distinguish one NGN entity from another. The NGN entity may be an endpoint, such as a telephone, or it may be a service delivery agent, such as a service provider (ETSI TR 187 010; ETSI 2008b). Telecommunications standards do not define identity, but they do define identifiers. Identifiers may be a series of digits, characters, and symbols used to uniquely identify subscriber, users, network elements, functions, or network entities providing services and applications. Identifiers fall into any of the three classes defined in (ETSI TS 184 002; ETSI 2006b):

1. Those generated automatically by network elements (e.g., call identifiers)
2. Those that may be allocated by operators without reference to external bodies (e.g., customer account number)
3. Those that are allocated to operators by external bodies (e.g., E.164 numbers, public IP addresses)

An overview of the NGN identifiers is provided in Table 16.1 (the list in the table should not be considered as providing a complete list).

The formats SIP Uniform Resource Indicator (URI) and TEL URI are a means to translate the identifiers used by the network to format E.164 numbers in coherence with SIP protocol requirements.

Table 16.1 Overview of Identifiers in NGN

USER/SERVICE IDENTIFIER	PUBLIC ID (USER AWARE)	FORMAT OF THE PUBLIC ID WITHIN THE NETWORK	PRIVATE ID (NETWORK AWARE)	NGN LAYER
User/Service Identifier	Name(s)	SIP URI	ID stored in ISIM	
	Number(s)	TEL URI; SIP URI with domain operator —provided	ID stored in ISIM or derived from USIM	Service
Network Identifier	Address	Number, and Routing Number; IP Address	Network ID; Line ID	Transport

Identifiers for Users

An NGN operator can store the user IDs either into the ISIM provided to the subscriber or directly inside the terminal if necessary. The ISIM itself is made up of various attributes. The main 3GPP attributes used for registration and authentication are the Home Domain Name, the Public Identifier, and the Private Identifier.

The Home Domain Name identifies the home domain of the user. It is used during authentication and registration. The format of the Home Domain Name is based on the Domain Name, for example, *operator.com* as specified in RFC 1035 (Mockapetris 1987b). The Home Network Domain Name is the parameter used to route the initial SIP registration requests to the home operator's IMS network. The Home Network Domain Name is stored in the ISIM. Upon receipt of the register information flow, the Proxy-Call Session Control Function (P-CSCF) examines the Home Network Domain Name to discover the entry point to the home network (i.e., the I-CSCF). For the initial registration message routing, the Domain Name resolution mechanism is made available to the user equipment at network attachment when the Dynamic Host Configuration Protocol (DHCP) provides it with the domain name of a P-CSCF and the address of a DNS capable of resolving that P-CSCF name. If there is no ISIM application (i.e., when there is no IMS-specific module), the Home Domain Name must be derived from the data available locally to the user entities (UEs).

Every NGN user has at least one private identifier. A private user identifier is assigned by the home operator and is used to identify the IMS user's subscription. Its main role is to support the authentication procedure during registration, reregistration, and deregistration; authorization; administration; and accounting at the home IMS. It is also used as the primary means of identifying the user within a dialogue between network entities (e.g., when fetching profile attributes from the User Profile Server Function (UPSF) or when selecting the S-CSCF). The private user identifier is not used in SIP call routing but is conveyed in all registration requests. A private user identifier is permanent* and is stored locally in the ISIM that will be used to instantiate the registration message parameters. In some cases, the private user identifier

* That is, not tied to a particular call instance or session.

may also be instantiated with default values when no ISIM is available. For its syntax, the private user identifier takes the form of an Network Access Identifier (NAI) and has the form username@realm. If there is no ISIM on the UICC, the private user ID is derived from the IMSI. The username is replaced with the complete IMSI value. For the *realm*, the Home Domain Name value is used. In IMS, the Home Domain Name, which is part of the private identifier, is used to route from a P-CSCF to the home operator's I-CSCF.

Every IMS user has one or more public identifiers, which are primarily used for user-to-user communication. The public identifier serves as a basis for message routing, possibly after a translation mechanism when appropriate, both for IMS session-based SIP messages (e.g., INVITE) and off-session SIP messages (e.g., NOTIFY). There is at least one public identifier stored in the ISIM, but, like the private identifier, in some cases, it may also be instantiated with default values when no such ISIM is available. Public identifiers are not authenticated during registration, but the correspondence between private identifier and public identifier can be checked by the UPSF, which is responsible for holding the user-related information, such as service-level user identification, numbering and addressing information, user security information, user location information, and user profile information (ETSI ES 282 001; ETSI 2005).

For its syntax, the public identifier takes the form of either an SIP URI or a TEL URI. An SIP URI takes the form sip:user@domain. TEL URI public user identifiers (whether they are a public E.164 [ITU 2010] number or a private number) cannot be used for SIP call routing in IMS and must be translated into SIP URIs using ENUM (described later). A public user ID, stored on the ISIM, for example, is needed by the NGN to access ordinary services (e.g., email, instant messages). However, access to emergency services may be possible without having a public user ID.

The IMS Public User ID (IMPU) is in the format of an SIP URI. However, both Internet naming and telephone numbering is supported, so technically the format is either an SIP URI or a TEL URI. At least one IMPU is stored on the ISIM and cannot be modified by the user equipment (UE). However, additional IMPUs do not have to be stored on the ISIM, although it is recommended. The IMPU can be registered either explicitly or implicitly[*] but must be registered before the identifier can be used to originate IMS sessions and IMS session unrelated procedures. The identifier also has to be registered, either explicitly or implicitly, before terminating IMS sessions. If a UICC is used that does not contain an ISIM, then a temporary IMPU can be derived from the USIM's IMSI and used for the initial SIP registration process only. The temporary IMPU takes the format of an SIP URI. The format is as follows:

```
SIP:<private user identifier>
```

where <private user identifier> is the private user identifier described in the previous paragraph.

The UPSF provides a public user ID to the user that will be used in the subsequent messages in the FROM field of the SIP INVITE message. As a consequence, the temporary IMPU is never displayed at the called party's UE.

Public user identities may be shared across multiple private user identities within the same IMS subscription. Hence, a particular public user identity may be simultaneously registered from multiple UEs that use different private user identities and different contact addresses. If a public user identity is shared among the private user identities of a subscription, then it is assumed that all private user identities in the IMS subscription share the public user identity.

The access up to the UE (user entity) is identified using the following:

Identifier for the access network
Identifier for the termination point of the physical transport on the access network (e.g., access link)
Identifier(s) for the logical channel (possibly recursive)
Identifier(s) for the UE using the same logical channel

Access network level registration involves access authentication, which is an authentication and authorization procedure between

[*] When a user has a set of public user identities defined to be implicitly registered via single IMS registration of one of the public user identity's in that set, it is considered to be an implicit registration. Nevertheless, the IMS supports implicit and explicit registration of further public user identities.

the UE and the Network Attachment Subsystem (NASS) to control the access to the access network. Two authentication types are considered for access networks: implicit authentication and explicit authentication.

Explicit authentication is an authentication procedure that is explicitly conducted between the UE and the NASS. It requires a signaling procedure to be performed between the UE and the NASS.

Implicit authentication does not require the NASS to explicitly conduct an authentication procedure directly with the UE; however, the NASS performs the implicit authentication based on identification of the layer 2 (L2) connection to which the UE is connected.

Authentication between users–subscribers and application–service providers may be explicit or implicit (based on trust–security assertions).

Through the NASS it is possible to

- Provide registration at access level and initialization of user equipment for accessing the NGN services
- Provide dynamic provisioning of IP address and other user equipment configuration parameters (e.g., using Dynamic Host Configuration Protocol)
- Authenticate the user, prior to or during the IP address allocation procedure
- Authorize of network access, based on user profile
- Access network configuration, based on user profile
- Enable location management

The Network Access Configuration Function (NACF) is a functional entity of NASS responsible for the IP address allocation to the UE (ETSI EG 201 940; ETSI 2001). It may also distribute other network configuration parameters such as the address of DNS servers, the address of signaling proxies for specific protocols (e.g., the address of the P-CSCF when accessing the IMS). The NACF provides an access network identifier to the UE. This information uniquely identifies the access network to which the UE is attached.

The Connectivity Session Location and Repository Function (CLF) is a functional entity of NASS that is responsible for registering the association between the IP address allocated to the UE and related network location information provided by the NACF, for example, access transport equipment characteristics and line identifier logical access ID. The CLF interfaces with the NACF to get the association between the IP address allocated by the NACF to the end-user equipment and the line ID. The CLF responds to location queries from service control subsystems and applications. The actual information delivered by the CLF may take various forms (e.g., network location, geographical coordinates, post mail address), depending on agreements with the requestor and on user preferences regarding the privacy of its location. The CLF holds a number of records representing active sessions.

For identifying the network nodes, the CSCF, Border Gateway Control Function (BGCF), and Media Gateway Control Function (MGCF) nodes—Interconnection Border Control Function (IBCF) and Interconnection Border Control Function (IBGF) functionalities—shall be identifiable using a valid SIP URI (host domain name or network address) on those interfaces supporting the SIP protocol (e.g., Gm, Mw, Mm, Mg reference points defined in ETSI ES 282 001; ETSI 2005).

These SIP URIs would be used when identifying these nodes in header fields of SIP messages. The names are allocated in the public DNS system; however, this does not require that the nodes be reachable from the global Internet. These URIs will not be resolvable via the public DNS, they will resolve only from within the operators' network. Globally unique identifiers for certain network elements (e.g., x-CSCFs) will be required so that a shared interconnect model, such as a GRX*/IPX† type interconnect model, can be supported. Element identifiers can be left to the choice of the service provider since the operator identifier and root domain uniquely identify the service pro-

* The GPRS Roaming eXchange (GRX) was implemented to support data roaming and authentication capabilities between GSM networks. It also supports 3G and wireless local area networks (LANs) and functions as a private network that connects GSM operators with a "best effort" approach to allow roaming customers access to the services offered by their home networks. The GRX establishes the interconnection for mobile data services along with some other applications like SMS and MMS.

† The eXchange (IPX) builds on the GRX concept by incorporating connectivity to non-GSM service providers such as fixed-line operators and application providers. It also leverages the infrastructure and experience with circuit-switched and packet-switched interconnects as well as the conversion being done today between the two technology domains.

vider. However, the element name should be compliant with (ITU-T Q.708; ITU 1999), and it is possible that further constraints, yet to be identified, may be required.

Identifiers for Services

Services provided by NGN also have to be identified. All public service identifiers need to meet the specific requirements of services such as voice, instant messaging service, presence service, and location service. The public service identifier shall take the form of either an SIP URI or a TEL URI.

A public service identifier defines a service or a specific resource created for a service. The domain part is predefined by the NGN operators, and the IMS system provides the flexibility to dynamically create the user part of the public service identifiers (PSIs). The SIP URI takes the form of a distinct PSI `sip:service@domain`, where `service` identifies a service. For example, a conference service at abinstitute.com may have the following SIP URI:

```
sip:conference@abinstitute.com
```

Universal Communications Identifier

One identification scheme that has been described outside the scope of NGNs is the Universal Communications Identifier (UCI; ETSI EG 201 940; ETSI 2001 and ETSI EG 202 067; ETSI 2002). The ways UCI could be implemented within an NGN context, however, are described in ETSI EG 203 072 (ETSI 2003).

Most identifiers are designed with a fairly simple set of requirements that relate to the need to provide a scalable, structured means to uniquely identify a resource within a narrowly defined technical framework such as a service or network infrastructure (e.g., the IP address within the Internet, and the E.164 [ITU 2010] telephone number within the telephone network). Dependent on the technologies, these identifiers might reference a diverse range of resources such as information objects, network endpoints, and service delivery agents such as a service provider. The focus of most identifiers is thus, at least initially, framed in terms of the capabilities and limitations of that technology.

The UCI is designed to meet a much broader and more complex set of requirements related to the simple human desire to be able to communicate in a way that allows both parties to understand and trust the identity of the other party. The frame of reference for the UCI is therefore primarily related to the communication needs of humans rather than to the technical capabilities and limitations of specific technology platforms. An explanation of the UCI requirements is given in the following paragraphs, prior to a description of the way the UCI design allows those requirements to be met.

Instead of being related to entities within the underlying communications technologies, the UCI is proposed as a means to identify people (or business roles) in the widest possible range of communications contexts, independent of the actual communication platforms and services used to enable the communication. The universality implied by the term *universal* relates to the potential applicability across all current and future means of communication, offering the same absolute global applicability that currently applies to telephone numbers and IP addresses.

An important part of the UCI concept is that the systems and infrastructure directly supporting UCI are intended to interpret and act on the wishes of the people initiating and receiving a communication. To make the rather abstract concept of acting upon the user's wishes concrete, a Personal User Agent (PUA) is introduced as part of the UCI support environment (ETSI EG 202 067; ETSI 2002). The PUA has access to a user profile that contains a wide range of user preferences that relate to the user's communication preferences across a wide range of contexts. Separate preferences may apply to each specific context, and the contexts can differ according to a number of situational variables such as time, place, and the presence of others.

The wide range of user preferences that could be contained in a user profile is outside the scope of this book but can be found in ETSI ES 202 746 (ETSI 2009). To ensure that the wishes of both parties to a communication are met, the PUA of the originating party may engage in a dialogue with the PUA of the party being contacted in order to try to meet the preferences of both parties and to take into account the technical capabilities of the communications devices and services that each of the two parties possess.

A key requirement behind the UCI is that it has to meet a number of strict and very different criteria in its role as an identifier. The first criterion is that the identifier should be globally unique. Uniqueness is essential in any identification system that is intended to identify specific individual resources. Ensuring that the scope of the uniqueness is global is also essential to ensure that the UCI communication is not restricted by any national or organizational boundaries. In addition, a person using a UCI should be able to contact other people who do not have a UCI, and vice versa.

All of these first three criteria are basic for ensuring that person-to-person communication using UCI can be relied upon irrespective of whether the communication is to, from, or between UCI users wherever they may be. An important aspect of an identifier from a human perspective is that it should be stable. The concept of stability implies that the same identifier can, if desired, be used throughout the lifetime of the user irrespective of any changes that the user may make to the communications devices and services that they choose to use. When this stability criterion is applied to UCI, it is beneficial to all UCI users as it means that once someone knows another person's UCI that same UCI will work many years later. The stability criterion cannot be guaranteed for telephone numbers, which may change if a national numbering plan is changed or if a person does not, or cannot, port their existing telephone number to a new service provider, nor does it apply to email addresses, which may fail to work or fail to be read if an email user moves to a different email provider.

Another very important aspect of person-to-person communication is the degree to which the identity of the other person (or role) is recognizable. Many identifiers that are currently used in person-to-person communication do not, of themselves, provide very meaningful indications of the identity of the person at the other end of the communication. The poor mapping of the identifiers is usually to do with the inherent constraints of the identification scheme. For example, the numeric format of the E.164 (ITU 2010) makes it impossible for the identifier to convey human-meaningful information such as personal names. Also, people will often be prevented from using their preferred name in the username part of an email address as there may be many people who all wish to use the same name, which cannot be allowed as it would break the need to preserve uniqueness of this element of an email address within any one domain.

The final important aspect of an identifier from a human perspective is whether the person receiving the communication can trust that when the identifier appears to include the true identity of the other party, that claim of identity authenticity can be believed. A large percentage of fraud, both over the Internet and via telephone calls, arises from the person initiating the communication claiming to be someone, or have some role, that is not his or her true name or role.

ETSI EG 201 940 (ETSI 2001) gives a detailed analysis about how this set of requirements can be met by an identifier. It concludes that no simple identifier can meet these requirements and that the only solution is a multipart identifier composed of three parts:

The label part: an alphanumeric label that allows the user to provide a string of human readable text that is the name by which they wish to be known for a specific communication. This part of the UCI is provided primarily to meet the human requirements of meaningfulness and also contributes toward the trustworthiness requirement.

A numeric part: a globally unique identifier in the form of an E.164 (ITU 2010) number, which is used to uniquely identify a specific user or role and the core part of the UCI that is processed during the establishment of all communications that use UCIs.

An additional information field: contains information that can serve a number of purposes including qualifying the label (e.g., indicating whether the label being used claims to be an alias or an authentic identity); providing the PUA at the other end of a communication with information that may assist it to propose the most mutually agreeable communication option; and enabling a person who has stored the UCI together with its additional information field to determine how best to communicate with a UCI user in future communications (by providing relevant information directly to a user who is about to communicate).

A single user may choose to use a range of different labels that he wishes to have presented to the person with whom he is communicating, varying which one is used either automatically (by rule) or choosing it manually according to contextual factors such as whether the other person is a friend, business colleague, or someone totally unknown. Any one of the labels can be either an *alias* or an *authentic* label. When the labels are used, the other party (or more precisely the other party's PUA) knows whether the label is an alias or authentic label according to a flag that is set in the additional information field. When an authentic label is used, the user is claiming that she is entitled to use the given label text and that this text gives a meaningful indication of her true identity. No such claims are being made for an alias label, which makes an alias label much closer to the identity labels contained in the email addresses and instant messaging identifiers that are more commonly used today. UCI does also support the concept of the right to communicate with anonymity by allowing the label field to be a null string.

EG 203 072 (ETSI 2003) gives some examples of the range of information that could be included in the additional information field that are all targeted at helping the users to optimize the communication to their particular preferences, including needs related to disabilities. However, the additional information that is of prime significance from the identifier perspective is the flag that specifies whether the label claims to be authentic or not. Where authenticity is claimed, the validity of that claim would have to be checked by the PUA and could use a number of trustworthy identity, credentialing, and access management methods to validate the authenticity of the claim.

The final element of the UCI is the numeric part, which is an E.164 (ITU 2010) number. The E.164 number is chosen because it is the only solution ensuring that the UCI could be globally reachable from all technologies, including the most basic public switched telephone networks, using very basic telephone terminals (ETSI EG 201 940; ETSI 2001 and ETSI EG 203 072; ETSI 2003).

E.164 Number Mapping (ENUM)

ENUM, which is defined in RFC 6116 (Bradner et al. 2011), is a means of resolving E.164 (ITU 2010) numbers into URIs by means of data stored in NAPTR resource records within the DNS. Processing of the

NAPTR records allows a possible range of services associated with the E.164 number to be identified.

RFC 6116 (Bradner et al. 2011) defines a procedure for converting a fully qualified E.164 (ITU 2010) number into a Fully Qualified Domain Name (FQDN).* A DNS query using this FQDN is used to return Name Authority Pointer (NAPTR) records, and the returned records are then processed according to specified methods.

The method for converting from an E.164 (ITU 2010) number to the FQDN is a multistep process as follows:

1. Ensure that the E.164 (ITU 2010) number is written in its full form, including the country code. Example:

    ```
    +44-7700-900123
    ```

2. Remove all nondigit characters with the exception of the leading "+":

    ```
    +447700900123
    ```

3. Remove all characters with the exception of the digits:

    ```
    447700900123
    ```

4. Put dots (".") between each digit:

    ```
    4.4.7.7.0.0.9.0.0.1.2.3
    ```

5. Reverse the order of the digits:

    ```
    3.2.1.0.0.9.0.0.7.7.4.4
    ```

6. Append a top-level domain name to the end:

    ```
    3.2.1.0.0.9.0.0.7.7.4.4.e164.arpa
    ```

RFC 6116 (Bradner et al. 2011) identifies the top-level domain of `e164.arpa` for the main public implementation of ENUM.

* An FQDN is the complete domain name for a specific computer, or host, on the Internet. The FQDN consists of two parts: the hostname and the domain name.

Use of public ENUM is not directly supported within the NGN specifications. However, in ETSI TS 184 010 (ETSI 2011c) the use of ENUM and DNS principles in the implementation of an interoperator backbone network is described. This implementation, which can be used for routing and to support a range of interoperator operations, is based on the use of a private DNS with no interconnection to the public Internet. The use of a private DNS is based on the need to establish a trusted network in which it is possible for NAPTR records to be populated only by recognized communications service providers. This separation from the public Internet is the only reliable way of ensuring that amendments to the ENUM database can be performed only by the operators responsible for that specific set of number(s), such as numbers for which they are the carrier of record. Although ETSI TS 184 010 describes a typical ENUM DNS hierarchy, it is not proscriptive in the use of this model. In some circumstances, where the DNS or ENUM database or parts of them are under a single administrative control, there is no need for operating dedicated root, for example, top-level domain, second-level domain, third-level domain. DNS server, respectively, that is, Tier-0, Tier-1, and Tier-2 ENUM server. In such a case, the whole DNS or ENUM data can be stored on one single device. This approach is called a nonroot DNS architecture. Because there is no name server hierarchy, a DNS/ENUM client can send DNS/ENUM queries directly to such a nonroot DNS/ENUM server. The message flow is optimized; there is just one DNS query and response message needed for DNS resolution.

17

Generic Authentication Architecture and Generic Bootstrapping Architecture

Nowadays, the users of fixed and mobile devices are able to use a huge amount of different services, most of them requiring some kind of authentication. Until now, the user had to have credentials for each service he aimed to use. Either he had to use them by hand, such as by writing his username and password when configuring them into a client application, or they were encased in a smart card, such as the Subscriber Identity Module (SIM) cards in mobile phones. The existence of several credentials is a problem due not only to its inconveniency for the user but also because provisioning these credentials to the users is expensive for the operators and other service providers (3GPP TR 33.919; 2011). For example, a third-party World Wide Web (WWW)-based service provider usually sends credentials via an email to the user and asks him to change the password. This, of course, is not very secure, and requires the user to choose an appropriate password and also to remember it. An example about the costs of provisioning for the operators is the distribution of SIM cards: they have to be manufactured and sent to the clients. They also have to be replaced when new services emerge and require a new kind of functionality.

Generic Authentication Architecture

Generic Authentication Architecture (GAA) provides a solution to the growing need for authentication and key agreement between the client and the services in the Internet or in the mobile network. GAA tackles the problem of authentication by using the already deployed and widely used Global System for Mobile Communications (GSM) authentication system as a basis for providing new credentials for both clients and servers. GAA is an authentication service provided by a mobile network operator, which allows the client and the service to

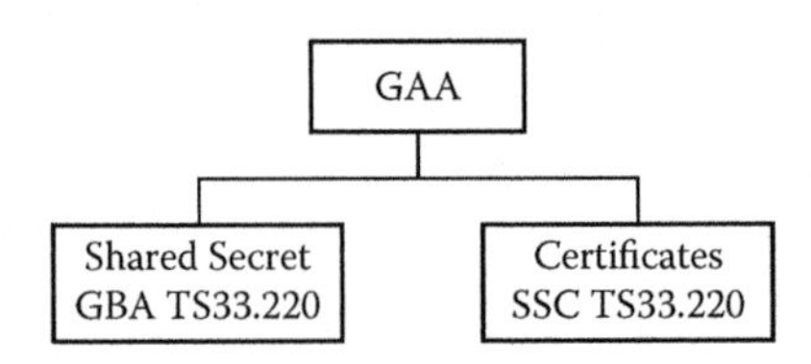

Figure 17.1 GAA schematic overview.

authenticate each other. However, GAA does not provide a Single Sign-On service (SSO),* just the shared secrets for the both parties.

There are two main ways of using GAA. The first is based on a preshared secret between the client and the server, and the second on public and private key pairs and digital certificates (Figure 17.1).

In the preshared secret case, the client and the operator are first mutually authenticated by means of the 3G Authentication and Key Agreement (AKA),† and they agree on session keys that can later be used between the client and the services the client wants to use. This is called bootstrapping. Generic Bootstrapping Architecture (GBA) that is described later can be used for this purpose. After that the services can fetch the session keys from the operator, and they can be used in some application-specific protocol between the client and the services. Several authentication protocols rely on a preshared secret between the two communicating entities, including Hypertext Transfer Protocol (HTTP) Digest (RFC 2617; Vixie 1999), Preshared Key Transport Layer Security (TLS; RFC 4279; Eronen and Tschofenig 2005), Internet Key Exchange (IKE; IETF RFC 2409; Harkins and Carrel 1998) with preshared secret and a priori any mechanism based on username and password.

In the second case, GAA is used to authenticate a certificate enrollment request by the client. First, the bootstrapping procedure is carried out as in the previous case. This assumes that the entity that needs to be authenticated (one or both partners in the communication) possesses a (public, private) key pair and a corresponding digital certificate that validates the key pair and binds the key pair to its legitimate owner.

* SSO enables users to access multiple services or system resources without having to provide authentication credentials more than once.

† Mechanisms describing how in a mobile context an AKA-based mechanism can be used to provide both communicating entities with a preshared secret are provided in 3GPP TR 33.919 (2011) and 3GPP TS 33.220 (2004c).

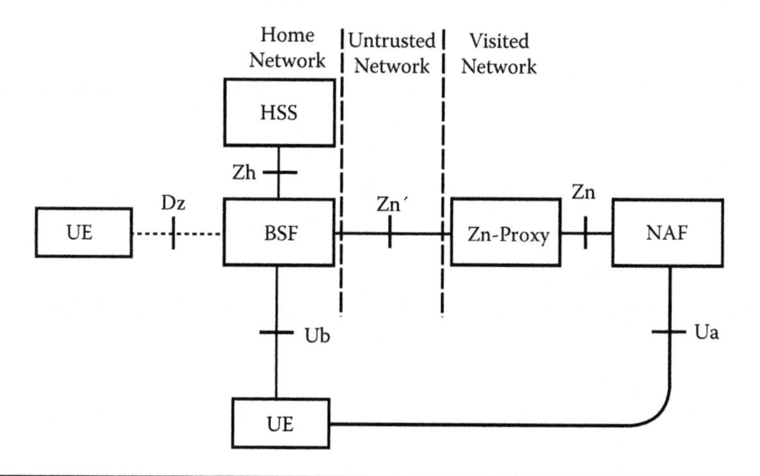

Figure 17.2 Simple network model for bootstrapping in visited networks involving HSS with Zh reference point.

Well-known protocols whose authentication is based on (public, private) key pairs include HTTP over TLS (RFC 2818; Rescorla 2000).

After this, the client can request certificates from the operator's Public Key Infrastructure (PKI),* where the authentication is done by the session keys obtained by accomplishing the bootstrapping procedure. These certificates and the corresponding key pairs can then be used to produce digital signatures or to authenticate to a server instead of using the session keys.

The main disadvantage of this type of authentication is that a PKI is needed and that asymmetric key cryptographic operations often require substantially more computational effort than symmetric key operations. Mechanisms describing how a mobile operator can issue digital certificates to its subscribers (hence providing a basic PKI) are provided in 3GPP TR 33.919 (2011) and 3GPP TS 33.221 (2004d).

Figure 17.2 shows the GAA network entities involved when the network application function is located in the visited network and the interfaces between them (3GPP TS 33.220 [2004c]). Optional entities are drawn with dashed lines and network borders with dotted dash. The user equipment (UE), for example, the user's mobile phone, and the Bootstrapping Server Function (BSF) mutually authenticate themselves over the Ub interface by using the HTTP Digest AKA (RFC 3310; Niemi, Arkko, and Torvinen 2002) protocol. The user entity

* PKI here means the use of public key and private key pairs for authentication.

also communicates with the Network Application Functions (NAFs), which are the application servers, over the Ua interface, which can use any application-specific protocol.

The BSF retrieves the subscriber's data (in the form of Attribute Value Pairs, AVPs) from the Home Subscriber Server (HSS) over the Zh interface, which uses the Diameter base protocol. If there are several HSSs in the network, BSF must first figure out which one to use. This can be done by either configuring a predefined HSS to BSF or by querying the Subscription Location Function (SLF)* over the Dz interface. NAFs retrieve the session keys from the BSF over the Zn interface, which also uses the Diameter base protocol. If the NAF is not located in the home network, it shall use a Zn-Proxy to communicate with BSF.

Generic Bootstrapping Architecture

GBA is described in 3GPP TS 33.220 (2004c). Its use can be divided into two procedures: the bootstrapping authentication procedure and the bootstrapping usage procedure. The bootstrapping authentication procedure includes the authentication of the client to the home network and deriving the key material. In the usage procedure the UE informs the NAF what key to use and the NAF then fetches this key from the BSF. There are also two different mechanisms for using GBA: Mobile Equipment-based GBA (GBA_ME) and GBA with UICC-based enhancements (GBA_U). The latter is a more secure one; however, it requires modifications of the Universal Integrated Circuit Card (UICC) and stores keys into the 3G Universal Subscriber Identity Module (USIM) instead of the 2G SIM application. In the following, the former mechanism is explained first, and then the differences between it and the GBA_U are provided.

Mobile Equipment-Based GBA

Before the UE starts to communicate with the NAF, it must figure out whether GBA is used. If the UE does not know this in advance, it starts the communication over the Ua interface without any GBA

* The SLF is a functional entity within an IMS that provides information about the HSS associated with a particular user profile.

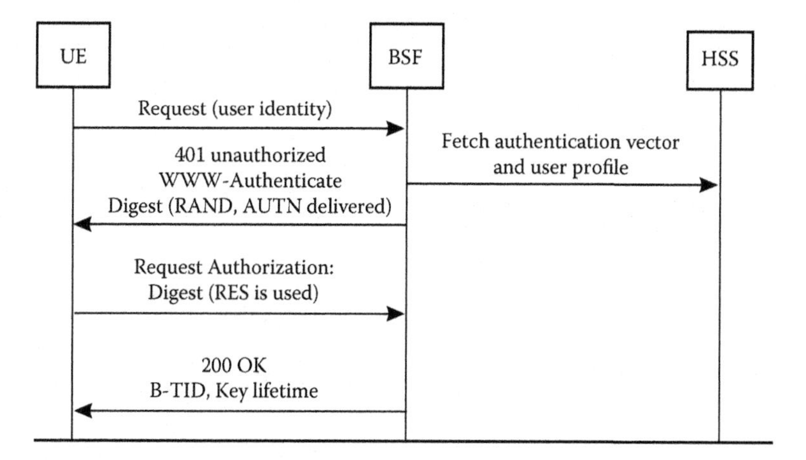

Figure 17.3 The bootstrapping authentication procedure.

parameters. If the NAF requires GBA, it responds with a request for bootstrapping initiation. The protocol used in the Ua interface is application specific. When the UE knows that the GBA is used, it performs a bootstrapping authentication to the BSF. Figure 17.3 shows the message flow chart of the procedure. The details of the used AKA authentication can be found in the 3GPP technical specification (3GPP TS 33.102; 2000), and the protocol uses a variant of the HTTP Digest (HTTP Digest AKA, RFC 3310; Niemi et al. 2002).

The authentication procedure consists of two request–response pairs between the UE and BSF. First, the UE sends a request with its username. The BSF then uses this username to fetch the corresponding GBA user security settings (GUSS) and an authentication vector from the HSS. The authentication vector consists of random number (RAND), authentication token (AUTN), expected response (XRES), and two cryptographic keys: cipher key (CK) and integrity key (IK). The BSF sends RAND and AUTN values with the authentication challenge response to the UE. Then the UE runs the AKA algorithms on its SIM, authenticates the BSF by verifying the AUTN, and derives the RES value and the session keys CK and IK. After this the UE sends the second request with the derived RES value. Then the BSF authenticates the user by comparing the RES from the UE with the XRES in the authentication vector. If they match, the UE is authenticated, and the BSF creates a Bootstrapping Transaction Identifier (B-TID) from the RAND value and the BFS's name. Then the B-TID is included in the 200 OK response message sent to the UE. In the

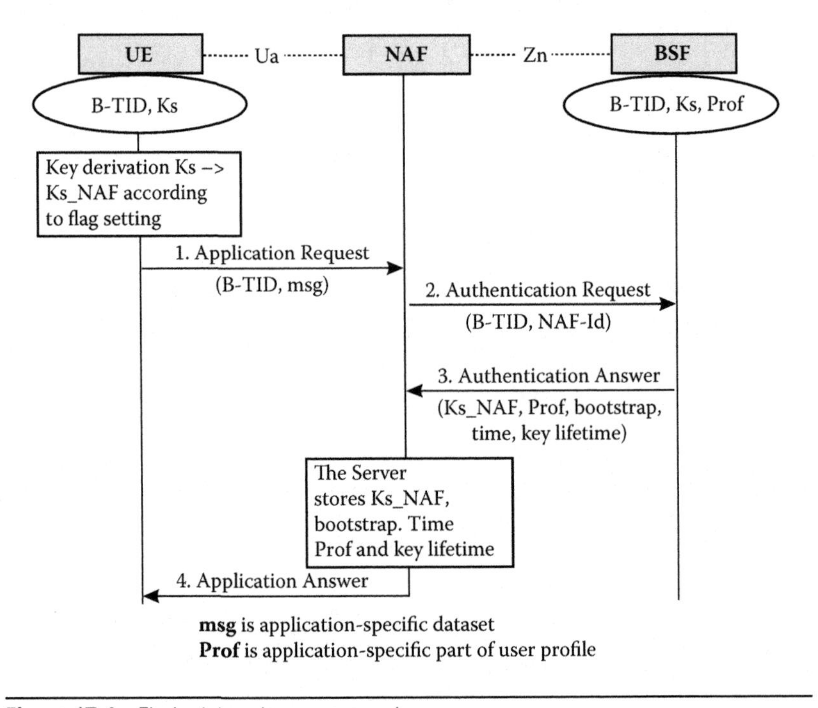

Figure 17.4 The bootstrapping usage procedure.

200 OK response the lifetime of the key material Ks is also included. Both the UE and the BSF are now able to create the Ks by concatenating the aforementioned keys CK and IK. However, the actual NAF specific key material (Ks_NAF) is computed from the Ks on demand, that is, in the UE when it starts communicating with the NAF, and in the BSF when the Ks_NAF is queried by the NAF (explained in the following). The Ks_NAF is created by using a key derivation function.

Figure 17.4 shows the message flow chart of the bootstrapping usage procedure. First, the UE sends an application request with the B-TID to the NAF. Then the NFA sends an authentication request to the BSF to get the key material corresponding to the given B-TID. The authentication request also includes the NAF-Id, which includes the NAF's public hostname used by the UE and the Ua security protocol identifier.* The BSF then verifies that the NAF is authorized to use the given hostname. If the verification is successful and there is a key found with the given

* Ua security protocol identifier is associated with a security protocol over the reference point Ua.

B-TID, the BSF sends the Ks_NAF with its bootstrapping time and lifetime to the NAF. In addition to the Ks_NAF, the NAF may also request some application-specific information from the BSF. However, if no key is found with the B-TID, the BSF tells this to the NAF, which then sends a bootstrapping renegotiation request to the UE. The UE must then perform the bootstrapping authentication procedure again.

GBA with UICC-Based Enhancements

In the GBA_U case, the idea is that all cryptographic operations are performed inside the UICC; thus, the keys CK and IK never leave this card. In the GBA_U bootstrapping procedure, the AUTN sent by the BSF in the 401 authentication challenge message is different from in the GBA_ME case. When the UE receives the challenge, the ME part of it sends the RAND and AUTN to the UICC, which then computes the CK, IK, and RES. Then the UICC stores the CK and IK values and gives the RES to the ME for sending it to the BSF. After this, the BSF and UICC are able to create the Ks_NAF as in the GBA_ME case. However, now two keys are created, Ks_ext_NAF and Ks_int_NAF. Both are created with the key derivation function: the former with the same parameters as the single key in the GBA_ME case and the latter with slightly different ones.

The usage procedure between the UE and the NAF is basically the same as in the GBA_ME case. The UE and NAF agree on which types of keys to use, the Ks_ext_NAF, Ks_int_NAF, or both. The NAF also tells the BSF if it supports GBA_U or not to get the suitable keys. The default key is always the Ks_ext_NAF to support UEs and NAFs that do not support the GBA_U. The bootstrapping usage procedure in the GBA_U case is depicted in Figure 17.5.

Support for Subscriber Certificates

The certificate support in GAA allows the issue of subscriber certificates for the UEs and delivering operator certification authority (CA) certificates. A subscriber certificate is a certificate issued to a mobile subscriber. It includes the subscriber's public key and optionally other subscriber's identity information.

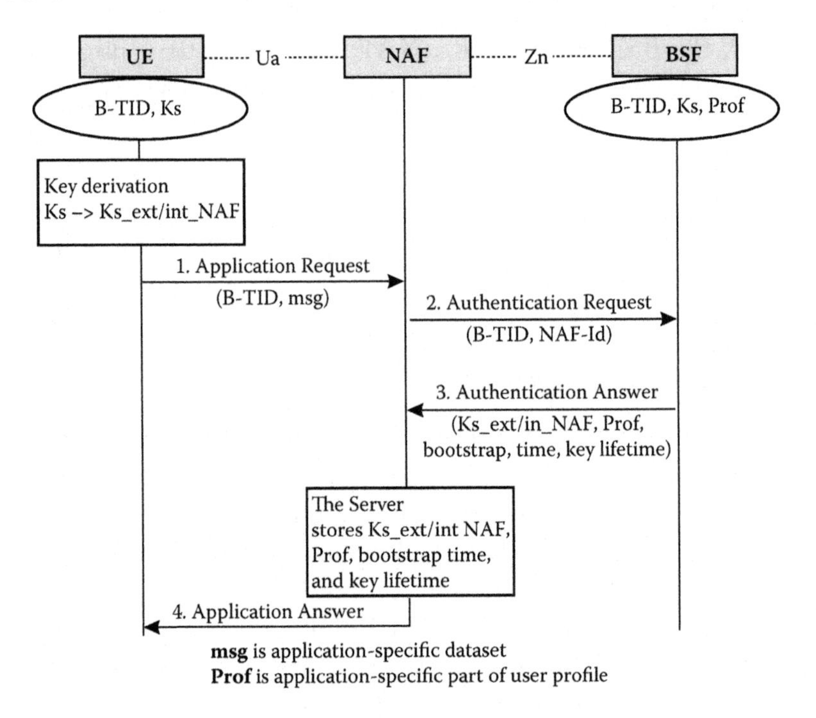

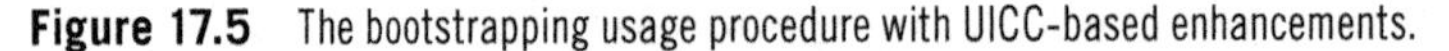

Figure 17.5 The bootstrapping usage procedure with UICC-based enhancements.

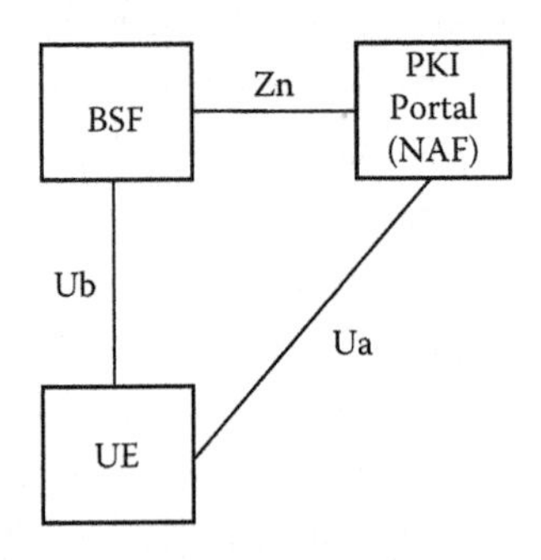

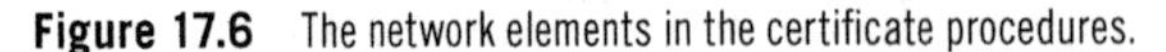

Figure 17.6 The network elements in the certificate procedures.

A CA certificate includes the public key of the CA. It uses the corresponding private key to sign the subscriber certificates. Figure 17.6 shows the elements involved in the certificate procedures. The PKI portal issues the certificates for the UE and delivers the operator CA certificate. The PKI portal is a PKI Registration Authority (RA), as it authenticates the certification requests from the UE. In addition, it can also function as a PKI CA and issue certificates. However, there might be an existing PKI infrastructure, which performs the role of

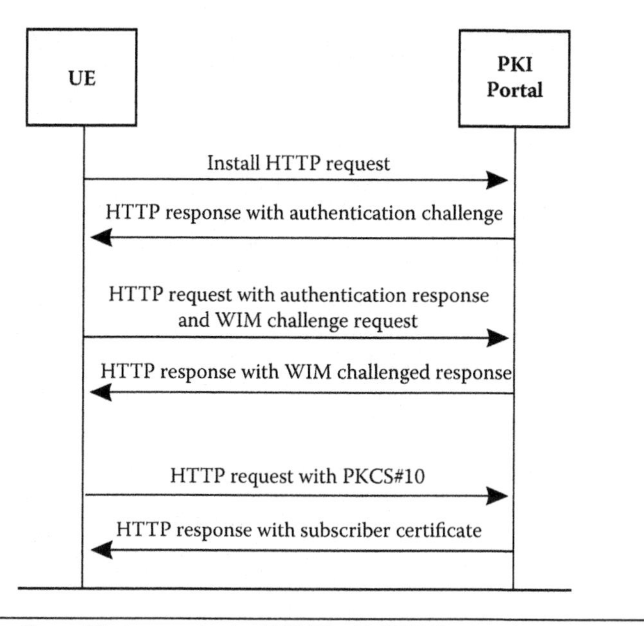

Figure 17.7 The certificate request using PKCS#10 with HTTP digest authentication.

a CA. BSF provides the key material for UE and the PKI portal as described previously and also the PKI portal-specific security data (e.g., the information of what kind of certificates the UE is allowed to enroll). The UE might have a WAP Identity Module (WIM) in which the private key is stored. The WIM is a tamper-resistant device, which is capable of providing proof that the key is actually securely stored in it. WIM could be, for example, a tamper-resistant SIM card.

Figure 17.7 shows the message flow chart of the certificate request. First, the UE sends an empty HTTP request to the PKI portal, to which the PKI portal responds with a Digest challenge. Then UE uses B-TID as a username and Ks_NAF as the password and generates a proper authorization header for the next request message. Next, if the certificate is requested for a WIM application in the UE, the UE creates a WIM challenge request that contains parameters needed for a key proof-of-origin generation. After receiving the request, the PKI portal fetches the corresponding Ks-NAF from the BSF, which stores the B-TID, to perform the normal HTTP Digest authentication procedure. If an extra assurance for the WIM application is needed, the PKI portal may use specific user security settings also fetched from the BSF to generate a WIM challenge response. After this the UE generates a Public-Key Cryptography Standards (PKCS#10) request

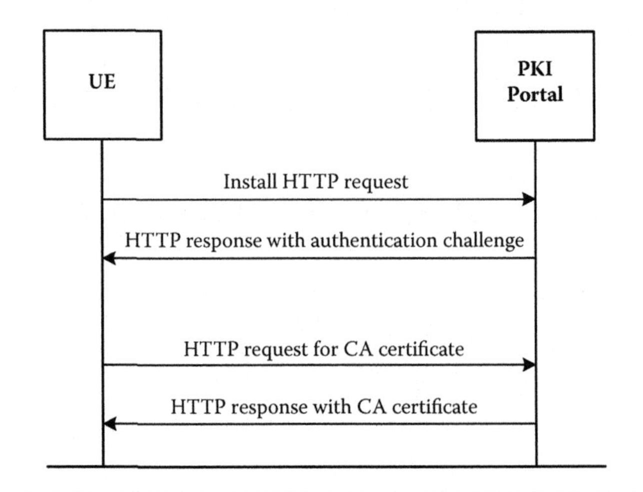

Figure 17.8 The CA certificate delivery with digest authentication.

(RSA Laboratories 2000). If a WIM application is used, its identifier and a proof that the key is stored are added to the request. If the PKI portal is not a CA, it forwards the request to a CA by using some available protocol. After the request has been processed, the new certificate is delivered to the PKI portal. The PKI portal will then send a response to the UE that includes either the certificate (or a pointer to it) or a full certificate chain.

Figure 17.8 shows the message flow chart of the CA certificate delivery. First, the UE starts with an empty HTTP request (just like in the subscriber certificate request procedure), and the BSF responds with a challenge. After receiving the challenge response, the UE creates a new request message with the CA issuer name in the request Uniform Resource Locator (URL). The digest authentication is performed as in the subscriber certificate request procedure. When the PKI portal receives the new request, it authenticates the UE and generates a response with the CA certificate. KS_NAF is used to authenticate and protect the integrity of the response by using the Authentication-Info header.* After receiving the response, the UE must first validate it. If the validation is OK, UE stores the certificate as a trusted CA certificate.

* The Authentication-Info header is used by the server to communicate information regarding the successful authentication in the response (RFC 2617; Franks et al. 1999).

X-3GPP Header Extensions

The use of Authentication Proxy (AP) is specified in 3GPP TS 33.222 (2004a). The AP in GAA is used to separate the GAA-specific authentication procedure and the Application Server (AS)–specific application logic to different logical entities. The AP is configured as a HTTP reverse proxy; that is, the fully qualified domain name (FQDN) of the AS is configured to the AP in such a way that the Internet Protocol (IP) traffic intended for the AS is directed to the AP by the network. The AP performs the GAA authentication of the UE. After the GAA authentication procedure has been successfully completed, the AP assumes the typical role of a reverse proxy; that is, the AP forwards HTTP requests originating from the UE to the correct AS and returns the corresponding HTTP responses from the AS to the originating UE.

The authentication of the UE is based on GAA. The AP removes the Authorization header from the HTTP requests that are forwarded from the UE to the AS and adds the Authentication-Info header to the HTTP responses that are forwarded to the UE from the AS. The UE may indicate the user identity intended to be used with the AS by adding a HTTP header to the outgoing HTTP requests. The HTTP header name is X-3GPP-Intended-Identity; the header contains the user identity surrounded by quotation marks.

The AP may be able to decide whether a particular subscriber, that is, the UE, is authorized to access a particular AS. The granularity of the authorization procedures is specified in 3GPP TS 33.222 (2004a). The AP may indicate an asserted identity or a list of identities to the AS by adding a HTTP header to the HTTP requests coming from the UE and forwarded to the AS. The HTTP header name is X-3GPP-Asserted-Identity, and it contains a list of identities, each of them surrounded by quotation marks and separated by a comma. In addition, the subscriber's application specific or AP-specific user security settings may be considered. The AP may indicate an authorization flag or a list of authorization flags from the application-specific User Security Settings (USS) to the AS by adding a HTTP header (X-3GPP-Authorization-Flags) to the HTTP requests coming from the UE and forwarded to the AS. 3GPP specific extension-headers for HTTP entity header fields are provided in 3GPP TS 24.109 (2004b).

18

Extensible Markup Language Document Management

The emergence of novel applications for the next-generation network (NGN) highlights the need to overtake the traditional "data silo" model, where the integration of different services is often performed vertically and per service. To fit this need, the Open Mobile Alliance (OMA) defined standard reusable common components called *enablers*. Enablers brought several advantages, such as the reduction of costs, a consistent definition of user interfaces across several services, and a uniform management of the increasing amount of user-related data. The last was facilitated by the gradual introduction of the Extensible Markup Language (XML) Document Management (XDM) technology.

XDM leverages on XML Configuration Access Protocol (XCAP) and Session Initiation Protocol (SIP) to allow services to access user-related information. In XDM, a user is identified by an XCAP user identifier (XUI), which takes the form of either an SIP Uniform Resource Identifier (URI) or a TEL URI. (The SIP URI is the preferred one in case the same user presents both identifiers.) User information is assumed to be stored in the form of a collection of XML documents residing in logical repositories, called XDM servers (XDMSs), which are specialized XCAP servers.

There are two different kinds of XDMSs. An enabler-specific XDMS is an XCAP server that allows a given *service enabler* (the terminology adopted by the OMA to refer to a generic service block) or one of its functional entity to manage XML documents for persistence purposes, for instance, the Resource List Server (RLS) used in OMA Presence SIP for Instant Messaging and Presence Leveraging Extension (SIMPLE).

An enabler is a functional entity that accepts and manages subscriptions to resource lists and distributes the resource state of the list to subscribers. RLS uses an RLS XDMS to store persistent XML documents containing information on the resource list state.

On the contrary, a shared XDMS is an XCAP server that stores common XML documents containing information that may be reused across various service enablers. The information contained in shared XDMSs, as in any other XCAP server is governed by its application usage. Early examples of common application usages in OMA included URI list and group usage list, both extending the *resource lists* application usage* described in RFC 4826 (Rosenberg 2007b). The URI list is intended to provide the capability to manage a simple list of URIs common to many enablers; a resource list within a RLS XDMS, for instance, may contain references to common URI lists stored in a shared-list XDMS. The group usage list defines a list of group names or service URIs whose type is a priori undefined but can be defined by specific applications. In XDM 2.0, shared XDMSs may run more complex application usages intended to provide facilities to manage groups of users, shared user profiles, and common-access policy rules governing communication requests.

XDM Aggregation Proxy

XDMSs are accessed not only by enabler-specific servers but also by XDM clients. Both an application server and end-user equipment can play the role of an XDM client. In the latter case, however, whenever the client is not in a trustworthy location, the access is not direct but is proxied through a functional entity called an aggregation proxy. The aggregation proxy is a Hypertext Transfer Protocol (HTTP) reverse proxy (i.e., a proxy server that hides the XDMSs from the client and acts on his or her behalf), which, after performing authentication of the XDM client and securing the connection using Transport Layer Security (TLS), routes XCAP requests to the correct XDMS. To perform authentication, the aggregation proxy uses the 3rd Generation Partnership Project (3GPP)–defined network authentication for Generic Authentication Architecture (GAA) or, alternatively, a weaker HTTP Digest authentication. The aggregation proxy

* A resource lists application is any application that needs access to a list of resources. The list of resources is a resource itself and is identified by a URI.

is responsible for providing the identity of the user to the XDMSs by inserting a special entity header* into XCAP requests.

In addition, the aggregation proxy may implement charging and compressing requests for clients with narrow bandwidth; however, these features are not mandatory in the XDM specification but are optional.

Each XDMS, being an XCAP server, conforms to the xcap-caps application usage, which consists in a single XML document, available in the *global* directory and listing the AUIDs, extensions, and namespaces understood by the server.

However, since there may be several XDMSs in a single network, it would not be very efficient for an XDM client to retrieve all xcap-caps documents directly from the XDMSs. To help the XDM client to discover the XCAP capabilities exposed by a network, the aggregation proxy collects the XCAP server capabilities from all the XDMSs, and upon client request, it conveys all the AUIDs, extensions, and namespaces merged into a single document.

One additional mandatory application usage defined by XDM is XML Documents Directory (`org.openmobilealliance.xcap-directory`). This application usage allows a client to retrieve a list of documents per given user collected either across all the XDMSs or from a given XDMS serving a specified AUID. Documents Directory maintains a managed XML document in each `user` tree branch named `directory.xml`. For each supported application usage (provided there are managed documents in the corresponding user tree branch), `directory.xml` lists information such as the Document selectors identifying the user documents, their ETag, their size, and the time of last modification. As for xcap-caps, the aggregation proxy acts as a collector and, upon request, serves to the client a single `directory.xml` (per user) merging all the entries in the different `directory.xml` documents of each available application usage.

* The header field name depends on the authentication procedure in use. In GAA, it could be a valid `X-3GPP-Intended-Identity` if provided, or, alternatively, it is an `X-3GPP-Asserted-Identity` provided by the aggregation proxy itself. In HTTP Digest authentication, the `X-XCAP-Asserted-Identity` entity header is used.

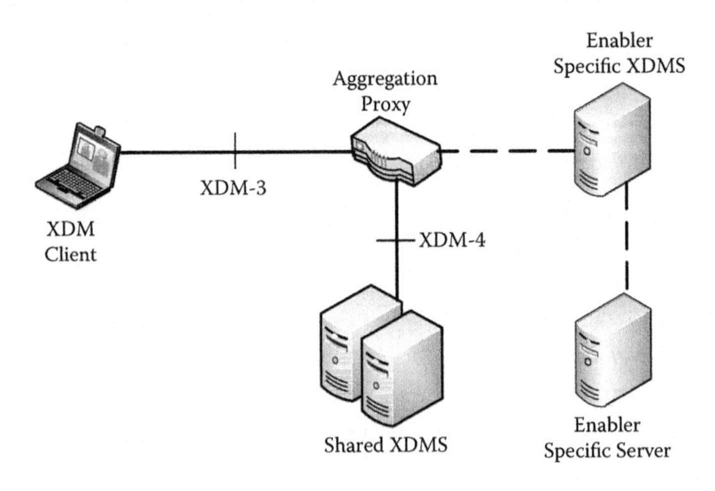

Figure 18.1 XDM architecture, functional entities, and reference points.

XDM Reference Points

The reference point between the XDM client and the aggregation proxy, for historical reasons explained later, is called XDM-3 and corresponds to the Ut reference point in the IMS (Figure 18.1). A second reference point is XDM-4, defined between the aggregation proxy and the shared XDMS. Both of them use the XCAP protocol.

Reference points XDM-1 and XDM-2 are no longer supported in the current release of the XDM standard (at the time of writing, XDM 1.1, released in June 2008).* Previously, they were used to support subscriptions to changes and notifications of changes in XML documents stored in XDMSs through the SIP protocol. Nevertheless, these capabilities have been reintroduced in XDM 2.0.

In particular, XDM-1 supports the communication between the XDM client and the SIP/IP Core network. In the IMS (Figure 14.5), when the XDM client is implemented in the user equipment, the XDM-1 reference point conforms to the Gm reference point. When it is implemented in an application server, XDM-1 conforms to the IMS Service Control (ISC) reference point. The XDM-2 reference point supports the communication between the XDMSs and the SIP/IP Core, and, in the IMS, it conforms to the ISC reference point.

* XDM 2.0 specifications, stable at the end of 2010, are not yet a standard but are a candidate standard.

Subscription and Notification Capabilities

Subscription to changes is performed by the XDM client sending an SIP SUBSCRIBE request. The simplest form of subscription is per single user and single application usage. In the SUBSCRIBE message, the Request-URI contains the target XUI with an additional parameter used to specify the AUID of interest (in the following example a shared group application):

```
SUBSCRIBE sip:joe@example.org;auid=org.openmobilealliance.groups
SIP/2.0
```

The client inserts a public identifier identifying the originating user agent in the P-Preferred-Identity header field (RFC 3325; Jennings, Peterson, and Watson 2002); in addition, it specifies the value `application/xcap-diff+xml` (which is an XML-based format hereafter described later) in the Accept header field. The Content-Encoding header field may indicate that the XDM client supports compression (e.g., the gzip algorithm). Finally, the body includes the XCAP resources to which the client is going to subscribe (in the following excerpt a group named *friends*):

```
<?xml version="1.0" encoding="UTF-8"?>
<resource-list xmlns="urn:ietf:params:xml:ns:resource-lists">
  <list>
    <entry uri="org.openmobilealliance.groups/users/sip:joe@
example.org/friends"/>
  </list>
</resource-list>
```

The SUBSCRIBE message is sent through the SIP/IP Core to the XDMS. The P-Preferred-Identity header field is replaced by the P-Asserted-Identity header field when the message is routed through the SIP/IP Core. The two fields are used within the trusted SIP infrastructure on behalf of the From header field to identify the user agent, making it possible to support Anonymous requests, where the From header field is set to *Anonymous.*

The XDMS, after verifying that the target resource exists and that the client is authorized to subscribe to changes (by default or by local policy rules), creates a subscription to changes of XCAP resources listed in the body of the SUBSCRIBE message and replies with an SIP 200 OK response message. Then, the XDMS generates an initial

SIP NOTIFY request message toward the subscribed XDM client. The message body contains an initial reference to XDM documents containing the resource for which changes have been subscribed. The format is compliant with the schema defined in (RFC 5874; Rosenberg and Urpalainen 2010) to represent changes in XCAP resources, which allows specifying the XCAP root URI, the interested document, its Etag, and its change set in terms of element or attributes.

```
<?xml version="1.0" encoding="UTF-8"?>
 <xcap-diff xmlns="urn:ietf:params:xml:ns:xcap-diff"
  xcap-root="http://xcap.example.com/"
  <document new-etag="18a22f"
   sel="org.openmobilealliance.groups/users/sip:joe@example.
org/friends"/>
 </xcap-diff>
```

Subsequent changes are likewise sent to the client using SIP NOTIFY request messages. The following excerpt reports that a change has occurred in the subscribed shared group document and a new entry has been added. Note that the value of the `previous-etag` attribute matches the value of the `new-etag` attribute of the previous request:

```
<?xml version="1.0" encoding="UTF-8"?>
 <xcap-diff xmlns="urn:ietf:params:xml:ns:xcap-diff"
  xmlns:list="urn:oma:xml:poc:list-service"
  xcap-root="http://xcap.example.com">
  <document previous-etag="18a22f"
 sel="org.openmobilealliance.groups/users/sip:joe@example.org/
friends"
    new-etag="937bc4">
     <change-log>
       <add sel="list:group/list:list-service/list:list">
         <list:entry list:uri="sip:alice@example.org">
       </add>
     </change-log>
   </document>
 </xcap-diff>
```

Subscription Proxy

XDM 2.0 allows for more complex forms of subscriptions; the same client can subscribe to changes occurring in more than one application and for more than one XUI very efficiently, using a single subscription request. This is realized by means of a subscription proxy to which the client subscription request is sent. The subscription proxy, upon receiving the subscription request, takes care of generating as many back-end subscriptions as the number of involved application usages and XUIs and collects changes from back-end XDMSes; changes are conveyed to the client using a multipart/related SIP NOTIFY request message whose body merges occurred changes in the subscribed resources (RFC 4662; Roach, Campbell and Rosenberg 2006).

Policy Rules

As in XCAP, the initial creator of a document is considered its primary principal, and as such, she is enabled to perform all operations on the document. It is not possible to assign permissions to manipulate a document to anyone except its primary principal or trusted applications. However, application usages may define additional rule sets to describe authorizations for accessing XCAP resources such as global documents. To facilitate this task, XDM borrows from RFC 4745 (Schulzrinne et al. 2007) the definitions of authorization policy rules through an XML-based policy framework.

Authorization policy rules are represented according to a simple schema and consist of three parts: conditions, actions, and transformations. The transformations part is obsolete and is ignored in XDM specifications. The actions part defines what should occur when the conditions are met and is extended by third parties according to their application-specific needs. The conditions part is standardized by XDM and allows specifying identities to be matched through their child elements: for example, a single user identity, multiple enumerated identities, all identities belonging to a given domain. XDM extends the original policy framework by providing few more cases: `external-list`, where the identity is taken from a list of identities stored externally and available through an XCAP URI (e.g., in a shared XDMS); `anonymous-request`, intended to match any incoming requests that have been identified as anonymous (as long as authenticated); and `other-identity`, which matches all identities not identified in any previous rules, allowing a default policy to be specified.

For instance, the following excerpt represents a rule set made of two rules containing, respectively, one and two conditions (the *actions* and *transformations* parts are empty).

```
<?xml version="1.0" encoding="UTF-8"?>
 <ruleset xmlns="urn:ietf:params:xml:ns:common-policy"
               xmlns:ocr="urn:oma:xml:xdm:common-policy">
     <rule id="20c1a">
         <conditions>
             <identity>
                 <one id="sip:joe@example.com"/>
                 <one id="tel:+1-212-555-1234"/>
             </identity>
         </conditions>
         <actions/>
         <transformations/>
      </rule>
      <rule id="20c1b">
         <conditions>
             <identity>
                 <many>
                     <except domain="example.org"/>
                     <except domain="example.net"/>
                     <except id="sip:alice@example.com"/>
                 </many>
             </identity>
         </conditions>
         <conditions>
             <ocr:external-list>
                 <ocr:entry anc="http://xcap.example.com/
resource-lists/users/sip:joe@example.com/index/~~/resource-lists/
list%5b@name=%22list_A%22%5d/">
                 <ocr:entry anc="http://xcap.example.com/
resource-lists/users/sip:joe@example.com/index/~~/resource-lists/
list%5b@name=%22list_B%22%5d/">
             </ocr:external-list>
         </conditions>
         <actions/>
         <transformations/>
     </rule>
  </ruleset>
```

The first rule exposes a condition matching two identities, one in the form of an SIP URI and the other of a TEL URI. The condition is evaluated to true whenever at least one of the listed identities is matched.

The second rule presents two conditions. The first condition identifies any authenticated identity (even anonymous) except specified user identities, which can also be expressed by referring to their domains. The condition is evaluated to true whenever none of the identity specified in the <except> element is matched. The second condition uses elements from the OMA extension of the policy framework, which allows for referral to identities stored in two external URI lists documents.

XDM 2.0 defines two additional children for the `<conditions>` element: `<media-list>` and `<service-list>`. The `<media-list>` element matches incoming requests associated with particular media types. Media types are defined as elements; some of them are noted in Table 18.1. The `<service-list>` element matches incoming requests associated with a particular service. Services are defined as <service> elements, each of which contains the `enabler` attribute that specifies a particular service enabler. Examples of values to be inserted into this attribute are given in Table 18.2. Each value corresponds to an enabler, which has associated a root namespace. The root namespaces, one for each enabler, are further divided according to enabler specific capabilities and specification version number, following the template:

```
urn:oma:xml:{enabler}:{SchemaSpecificTag}
```

Table 18.1 OMA-Defined Media Types to Be Used in Policy Rule Conditions

XML ELEMENT NAME	DESCRIPTION
<message-session>	Applications based on the Message Session Relay Protocol (MSRP), a protocol for transmitting a series of related instant messages in the context of an SIP session
<pager-mode-message>	Pager messaging: Instant Messaging applications that relies on SIP MESSAGE requests; contrary to message-session, each instant message stands alone
<file-transfer>	File transfer applications
<audio> <video>	Applications using audio and video capabilities
<poc-speech>	Push to Talk over Cellular speech (a walkie-talkie–like communication service implemented on top of the cellular network)
<group-advertisement>	Advertisements for recently created groups of users

Table 18.2 OMA-Defined Service Enablers to Be Used in Policy Rule Conditions

NAME	URN PREFIX	DESCRIPTION
Im	urn:oma:xml:im:	Instant Messaging
Mms	urn:oma:xml:mms:	Multimedia Messaging Service
Poc	urn:oma:xml:poc:	Push to Talk over Cellular
Prs	urn:oma:xml:prs:	Presence Service
Supm	urn:oma:xml:supm:	User Profile Management
Xdm	urn:oma:xml:xdm:	XML Document Management

These namespaces actually define XML schemas associated with each application usage.

Policy rules usually come from particular domains to meet specific application needs. For instance, Push-to-Talk over Cellular (a walkie-talkie–like communication service implemented on top of the cellular network) extends the policy framework by defining actions that govern the management of incoming invitations. However, with time some of these needs (e.g., control of incoming communication) have also become useful to many other applications. To accommodate the reuse of common access policy in a uniform way, XDM 2.0 has defined a shared application usage called *user access policy*, whose rules are contained in an XML document stored in a shared XDMS called shared policy XDMS.

Search Capabilities

XDM 2.0 allows a client to search information in any collection of XML documents stored in XDMS using a limited subset* of the XQuery functions.

Both the search query and the results are transported using the HTTP protocol. To issue a query, the XDM client sends a HTTP POST request to a search proxy server (or to the aggregation proxy if the client is not in a trustworthy location) implementing the `org.openmobilealliance.search` application usage. The HTTP request URI contains the following parameters as HTTP URI query parameters (e.g., as key-value pairs appended to the URI using a question mark and separated by the ampersand sign):

The target, that is, a parameter identifying the collections of document to search

The domain, an optional parameter allowing the specification of additional domains to search†

For instance, the following request URI

* Each application usage may support only some of the capabilities defined by the standard.

† As will be described, other than the home domain, it is possible to perform XDM operations (manipulations of, subscriptions to, and searches in documents) across different domains.

```
http://xcap.example.com/org.openmobilealliance.search?target=org.
openmobilealliance.user-profile/users/&domain=home%20example.org
```

instructs the search proxy to perform a search in the home domain (example.com) and in a remote domain (example.org).

XDM provides the facility of aggregating more than one query in a "search set," which is useful to reduce the overhead of exchanging multiple HTTP requests and response messages. The actual query is contained within the `query` element of each `search` entry in the search set encoded as character data. The `search` element also specifies a per-client unique query identifier (`id` attribute) and, optimally, an indication of the wanted maximum number of returned results (`max-results` attribute). The following excerpt illustrates a search set:

```
<?xml version="1.0" encoding="UTF-8"?>
<search-set xmlns="urn:oma:xml:xdm:search">
<search id="af200a" max-results="2">
  <request>
    <query>
   <![CDATA[
      xquery version "1.0";
      declare default element namespace "urn:oma:xml:poc:list-service";
      declare namespace rl = " urn:ietf:params:xml:ns:resource-lists";
      for $1 in collection("org.openmobilealliance.groups/users/
sip:joe@example.org/friends")/group/list-service
      where ($1/max-participant-count<4)
      and     ($1/list/entry[@rl:uri='sip:alice@example.org'])
      order by $1/max-participant-count descending
      return $1
  ]]>
    </query>
  </request>
</search>
</search-set>
```

This search set contains one query. The query defines a single variable `$1`, which iterates over elements (identified by the XPath expression/group/list-service) inside XML documents related to the application usage `urn:oma:xml:poc:list-service`. These documents are taken from the `friends` branch of the user joe@example.org. The convention here used to identify XML documents is similar to the one adopted by the XCAP document selector. However, unlike a traditional XCAP expression—which may contain one and only one document selector—the result of the XQuery `collection()` function is a sequence of documents, not one specific document.

The where clause defines conditions to filter out the elements; it is possible to define many conditions inside a single where clause using logical operators (in the previous example the logical operator "and" is used). The order by statement defines an order among the returned results, based on one or more variables' value.

The return statement, concluding the query, defines which results should be returned. In the previous example, just one collection of nodes is returned; however, there are also other possibilities, ranging from returning single values to whole XML documents.

Based on the AUID contained in the target parameter, the search proxy takes care of dispatching the query to the appropriate XDMS. This latter executes the query over all XML documents of the specified collections, and returns the results (aka query response) as an XML document contained in the body of the HTTP POST response message. The document consists of a search set similar to the one sent in the request message; unlike the latter, however, it contains response elements. The following excerpt illustrates a search set containing responses:

```
<?xml version="1.0" encoding="UTF-8"?>
<search-set xmlns="urn:oma:xml:xdm:search"
xmlns:list="urn:oma:xml:poc:list-service"
xmlns:rl="urn:ietf:params:xml:ns:resource-lists">
<search id="af200a">
 <response>
  <list:list-service uri="sip:tour1@example.org">
    <list:display-name lang="en">Tour-1</list:display-name>
    <list:list>
      <list:entry uri="sip:trudy@example.org">
        <rl:display-name lang="en">Trudy Smith</rl:display-name>
      </list:entry>
      <list:entry list:uri="sip:alice@example.org">
        <rl:display-name lang="en">Alice Doe</rl:display-name>
      </list:entry>
      <list:entry list:uri="sip:mark@example.org">
        <rl:display-name lang="en">Mark Lee</rl:display-name>
      </list:entry>
    </list:list>
    <list:max-participant-count>3</list:max-participant-count>
  </list:list-service>
  <list:list-service uri="sip:tour2@example.org">
    <list:display-name lang="en">Tour-2</list:display-name>
    <list:list>
      <list:entry uri="sip:alice@example.org">
        <rl:display-name lang="en">Alice Doe</rl:display-name>
```

```
      </list:entry>
      <list:entry uri="sip:joe@example.org">
        <rl:display-name lang="en">Joe Smith</rl:display-name>
      </list:entry>
    </list:list>
    <list:max-participant-count>2</list:max-participant-count>
  </list:list-service>
 </response>
</search>
</search-set>
```

The `id` attribute of the `search` element is used to associate each response to the corresponding query. The HTTP response message is eventually returned to the XDM client.

Communication with Remote Networks

XDM 2.0 introduces the capability to handle, subscribe, and search on XML documents hosted in XDMSs residing in a domain other than the home domain. This is achieved by means of two reference points:

1. IP-1 reference point, which conforms to the Ici reference point in the IMS, based on the SIP protocol (which is natively cross-domain). Subscriptions and notifications to remote XML document changes pass through IP-1.
2. NNI-1 reference point, supporting the HTTP-based communication between XDMSs in different domains via functional elements named cross-network proxies.

The cross-network proxy is a novel reverse proxy server introduced in the XDM 2.0 architecture. Communicating with its peer cross-network proxy in the foreign domain, it provides a uniform XCAP- and XQuery-based interface toward the XDM infrastructure in the remote network.

The local cross-network proxy accepts incoming requests from the local aggregation proxy (or from the local search proxy) targeted to a foreign domain. The target domain name is determined by inspecting the XUI of the XCAP URI in the outgoing request (or the domain parameter in the HTTP request URI in the case of a search request). The local cross-network proxy checks whether the target domain is a trusted domain, and if this is the case, it forwards the request to the remote cross-network proxy through the NN-1 reference point. The remote cross-network proxy, in turn, forwards the request to its local aggregation proxy (or search proxy).

Since all these elements are HTTP proxies, the response may simply come back to the client by following the reverse path.

SECTION V
Linked Data

The Resource Description Framework (RDF) was first introduced to provide machine-understandable description of digital artifacts on the network. It evolved along with the development of the concept of the Uniform Resource Identifier (URI) for any kind of resources, and it quickly become the standard descriptive language of the Semantic Web, providing the basis for other richer languages, such as the Web Ontology Language (OWL).

In 2004 Edward Feigenbaum, considered the "father of expert systems," made a breakthrough in how to think about semantics on the web: "Path of maximal return is more knowledge, not more logic" (Feigenbaum, 2004). Some years later, in January 2007 the Linking Open Data project, a community effort supported by the W3C Semantic Web Education and Outreach Group,* launched the nucleus of what was to become linked data.

Challenging problems are still unresolved, but during the last years linked data has attracted the attention of different stakeholders across various fields such as government, industry, academia, and research institutions, thus creating a vibrant community and boosting the development of an uniform standard for distributed data management on the web.

* The W3C Semantic Web Education and Outreach Group (SWEO), http://www.w3.org/2001/sw/sweo.

19
Resource Description Framework

Resource Description Framework (RDF) was first introduced to provide machine-understandable description of resources. While the term *resource* was originally intended to be synonymous with the generic "object of the network" (RFC 1630; Berners-Lee 1994), years of research, development, and evolution of the Semantic Web have imbued it with a different and broader meaning. The first extension of the notion of resource to "anything that has identity" occurred in August 1998, with the advent of RFC 2396 (Berners-Lee et al. 1998) and in January 2005 was broadened to include abstract concepts (RFC 3986; Berners-Lee et al. 2005). The history of RDF has thus been closely paired with the evolution of the concept of the Uniform Resource Identifier (URI) as a means to reference or identify *things*. Consequently, RDF, a language originally designed to describe digital artifacts in a machine-understandable fashion, has become the standard descriptive language of the Semantic Web.

RDF Triples

RDF uses URI references (URIRefs) to talk about things and their relationships. It distinguishes between individuals, kinds, properties, and values. Individuals, kinds, values that are not literal,* and properties as well, might be identified by URIRefs. The description is realized by means of statements in form of triples, or tuples containing three elements:

```
(s,p,o)
```

where *s* is the subject, *p* is the predicate, and *o* is the object. The degree of flexibility of this simple structure, and thus its expressivity, is quite elevated. In fact, the subject could refer to an individual, a class of

* Literals are simple strings of characters.

individuals, or even a property. There is currently debate as to whether to allow literals as subjects as well, but currently the standard does not allow for this. Thus, the subject is always in the form of a URIRef or a *blank node* (blank nodes represent local resources that cannot be accessed from outside a graph, as they lack a global identifier).

The predicate always refers to a property and is in form of a URIRef. The object, similarly to the subject, could refer to an individual, a class of individuals, or a property, in the form of a URIRef or a blank node; in addition, it could be a literal, and it may or may not have a type. Using a terminology later introduced by Web Ontology Language (OWL), the range of a property may consist of individuals or literals. Properties whose range is individuals are called *object properties*; the others are known as *literal properties*.

RDF statements are modeled as nodes and arcs in an RDF graph, which is a set of RDF triples with each subject or object of a RDF triple represented as a node. Each predicate is modeled as an arc originating from the node representing the subject to the node representing the object. Literals are modeled as terminal nodes and depicted as boxes. (Currently they cannot be the subject in any statement; thus they cannot have any outgoing arc.) Figure 19.1 illustrates an example of an RDF graph. The node on the far left of the picture represents a resource (in this example a person) identified by the URIRef <http://example.org/people#AliceSmith>. The arc connects this resource to a box that contains an RDF literal, which is a simple string. The arc represents the property identified by the URIRef <http://xmlns.com/foaf/0.1/name>, that is, name. The whole picture can be interpreted as "The name (<http://xmlns.com/foaf/0.1/name>) of the resource identified by the URIRef <http://example.org/people#AliceSmith> is Alice."

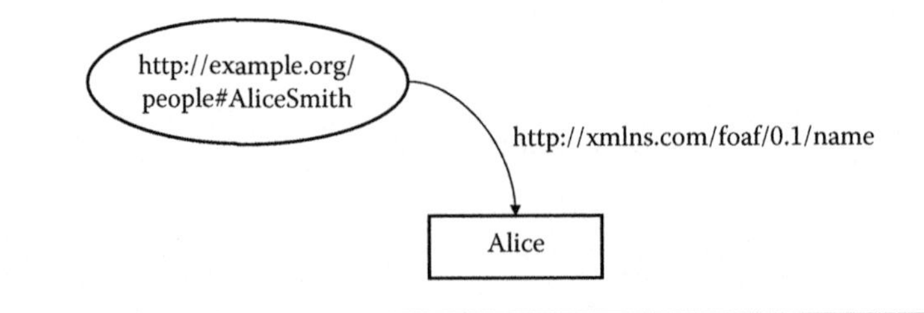

Figure 19.1 An RDF graph.

Apart from the graph model, RDF does not mandate a particular notation to represent its productions. Indeed, this makes it different from other languages, allowing RDF to be unbiased by serialization-specific constraints. In addition, this characteristic also allows the same RDF statements to be expressed using different notations (with some notations being better than others depending on the intended purpose). The primary serialization formats for RDF are RDF/Extensible Markup Language (XML), Turtle, and N-Triples. Other formats include RDFa, suitable for embedding RDF information within Hypertext Markup Language (HTML), and several JavaScript Object Notation (JSON)–based serialization formats that are still under discussion by the JSON community.

RDF/XML

RDF/XML was the first RDF serialization format standardized by the World Wide Web Consortium (W3C; Beckett 2004). RDF/XML encodes an RDF graph as a well-formed XML document. Compared with others, this notation can be perceived as complex, but its design attempts to provide the most fluid transition possible from the widely used XML documents to RDF documents. Most apparently, whenever allowed by the encoding rules, URIRefs are mapped directly to XML elements or attributes. To enable this, URIRefs are written as fully qualified names (FQNs). According to this encoding, a URIRef is split into a namespace (or prefix) and a localname and is encoded as XML elements or attributes identified by that FQN.* The original URIRef can be reconstructed by simply concatenating the prefix and the localname. This mechanism is evident when looking at the way RDF predicates are encoded. When the RDF predicate is a literal property, literals are directly represented as text enclosed by the XML element that represents the property. The literal's datatype, if any, appears as a value of the attribute `rdf:datatype` included in the element. The following example illustrates this usage:

```
<dcterms:date rdf:datatype="http://www.w3.org/2001/
XMLSchema#date">2010-10-12</dcterms:date>
```

* As in XML, some prefix names are well-known and commonly used in the practice. They are listed in Table 20.1 and Table 20.2.

Unfortunately, URIRefs appearing in attribute values are parsed literally as strings by the XML processor so they cannot be abbreviated using the FQN syntax; actually, in the previous example the URIRef http://www.w3.org/2001/XMLSchema#date was used instead of the more common FQN form `xsd:date`.

However, URIRefs in attribute values may be shortened using a syntax based on Document Type Definition (DTD) *entity references* defined in DTD. This syntax allows the XML processor to preserve a correct interpretation of the shortened URIRef. A DTD entity reference declaration is added at the beginning of the XML document to instruct the XML processor to perform the substitution:

```
<!DOCTYPE rdf:RDF [<!ENTITY xsd "http://www.w3.org/2001/
XMLSchema#">]>
```

This way the aforementioned line is turned into a more readable one:

```
<dcterms:date rdf:datatype = "&xsd;date">2010-10-12</dcterms:date>
```

When the RDF predicate is an object property, resources objects, identified by their URIRef, are usually encoded as a value of the attribute `rdf:resource`, as in the following example:

```
<dcterms:isPartOf rdf:resource = "http://example.org/
resources#aReview"/>
```

URIRefs representing subjects appear as a value of the `rdf:about` attribute inside the element `rdf:Description`.

```
<rdf:Description rdf:about = "http://example.org/
resources#aPaper">
```

The following example is a valid RDF/XML document describing an article (a resource) titled "The First Example Paper" (an untyped literal), which is part of a review (another resource) published on 2010-10-12 (a typed literal).

```
<!DOCTYPE rdf:RDF [<!ENTITY xsd "http://www.w3.org/2001/XMLSchema#">]>
<?xml version="1.0"?>
<rdf:RDF xmlns:rdf="http://www.w3.org/1999/02/22-rdf-syntax-ns#"
  xmlns:dcterms="http://purl.org/dc/elements/1.1/">
  <rdf:Description rdf:about="http://example.org/resources#aPaper">
    <dcterms:isPartOf rdf:resource="http://example.org/resources#aReview"/>
    <dcterms:title>The First Example Paper</dcterms:title>
```

```
  </rdf:Description>
  <rdf:Description rdf:about="http://example.org/resources#aReview">
    <dcterms:title>Examples</dcterms:title>
    <dcterms:date rdf:datatype="&xsd;date">2010-10-12</dcterms:date>
  </rdf:Description>
</rdf:RDF>
```

The DOCTYPE declaration appearing in the first line of the example allows the full namespace URIRef <http://www.w3.org/2001/XMLSchema#> to be abbreviated using the string &xsd; (in the previous example &xsd; occurs in the value of the rdf:datatype attribute).

If the URIRef representing the subject is a hash URI, the rdf:ID attribute can be used instead of rdf:about. Its value specifies the fragment identifier. By default, the fragment identifier is interpreted relative to the *base URI* which is assigned to the XML document. To change the base URI, the attributed xml:base is used.

Whenever known, the type of an individual can also be directly encoded as an element replacing the rdf:Description.

Finally, blank nodes are usually encoded through the rdf:nodeID attribute, containing the local identifier of the blank node (other possibilities exist however). The following example is similar to the previous one except that:

rdf:Description nodes have disappeared in favor of nodes describing the classes of the described resources.
xml:base has been introduced to declare the base namespace; resources identified by rdf:ID are part of this namespace.
One blank node is used (rdf:nodeID = "aReview").

```
<!DOCTYPE rdf:RDF [<!ENTITY xsd "http://www.w3.org/2001/XMLSchema#">]>
<?xml version="1.0"?>
<rdf:RDF xmlns:rdf="http://www.w3.org/1999/02/22-rdf-syntax-ns#"
  xmlns:dcterms="http://purl.org/dc/elements/1.1/"
  xmlns:class="http://example.org/class#"
  xml:base="http://example.org/resources">
  <class:Paper rdf:ID ="aPaper">
    <dcterms:isPartOf rdf:nodeID ="aReview"/>
    <dcterms:title>The First Example Paper</dcterms:title>
  </class:Paper>
  <class:Review rdf:nodeID ="aReview">
    <dcterms:title>Examples</dcterms:title>
    <dcterms:date rdf:datatype="&xsd;date">2010-10-12</dcterms:date>
  </class:Review>
</rdf:RDF>
```

Turtle

As previously mentioned, RDF/XML has been an explicitly declared attempt to try to attract the interest of the wide XML community by forcing RDF statements to be shaped within the more familiar XML constructs (as illustrated by the previous example). Forcing this fit has resulted in many rules that should or could be optionally followed when expressing RDF in XML. There are also further limitations such as the requirement, imposed by the XML syntax, that all RDF subjects must be URIRefs, thus preventing literals from being used as subjects. All of these characteristics, rules, and limitations have made this format less appealing for the RDF user community, which was instead oriented toward alternatives closely reflecting the simplicity of the RDF triple pattern. Turtle (Beckett and Berners-Lee 2008) was thus developed in response to the need for a simpler and more human-friendly textual serialization format. In Turtle, every URIRef is represented as an FQN, and prefixes are declared as *annotations* at the beginning of the Turtle document. As in XML, the choice of a prefix name is free and has relevance only in the context of the Turtle document. Blank nodes are also written in FQN form, with the prefix set to underscore and the suffix set to the local node identifier. RDF statements are simply written in the form of triples, one per line, each followed by a dot. Subject, predicate, and object in each triple and the dot itself is separated by spaces.

```
@prefix dcterms: <http://purl.org/dc/elements/1.1/> .
@prefix ex: <http://example.org/resources#> .

ex:aPaper dcterms:isPartOf ex:aReview .
```

Some shortcuts are allowed. Triples that share the same subjects can be abbreviated, omitting the subject in the lines following the first one and replacing the final dot with a semicolon. Similarly, triples that share the same subjects and the same predicate are expressed by omitting the subject and the predicate in the lines following the first one and by replacing the final dot with a comma.

Untyped RDF literals are simply enclosed in double quotes, whereas typed literals use the same syntax but the type specification is concatenated at the end of the closing quote, as in the following example (prefixes are omitted):

```
ex:aPaper dcterms:title "The Example Paper"^^xsd:string ;
```

Literals that are nouns in a given language are qualified by concatenating the language specification after the closing quote. For instance, *12* is not the same as `12^^xsd:integer`, and *apple* is a different literal from `apple@en`.

N-Triples

N-Triples (N3) is another notation that uses a syntax similar to Turtle; however, instead of using FQN, it uses full URIRefs enclosed in angled brackets. N3 leaves each line self-contained (and thus self-explanatory), at the cost of increased redundancy, as easily seen in the following example:

```
<http://example.org/resources#aPaper> <http://purl.org/dc/
elements/1.1/isPartOf> <http://example.org/resources#aReview> .
<http://example.org/resources#aPaper> <http://purl.org/dc/
elements/1.1/title> "The Example Paper"^^xsd:string .
```

The complete definition of the N-Triples syntax is described in Grant and Beckett (2004).

20
Advanced RDF

Resource Description Framework (RDF)–rich expressivity is achieved by means of an extendable set of terms grouped into vocabularies. However, the RDF core specifications include only a small number of general-purpose terms. Most RDF terms (including domain specific terms) are defined in third-party vocabularies by relevant standardization forums, organizations, or even individuals. Vocabularies define terms together with precise semantics for them. Users of a vocabulary are expected to understand the definitions and the semantics.*

RDF Schema and OWL

One of the most important vocabularies is the RDF Vocabulary Description Language (Brickley and Guha 2004), usually referred to as *RDF schema* (prefix `rdfs:`, namespace defined in Table 20.1), which enables RDF to describe kinds of things (i.e., classes and hierarchies). A semantically richer vocabulary is provided by the Web Ontology Language (OWL; McGuinness and van Harmelen 2004), which is built on an RDF core and an RDF schema.† OWL is partly derived from earlier attempts such as DAML+OIL (Connolly et al. 2001).

* Unfortunately, this is not always the case. A very well-known related issue is about the use of `owl:sameAs` property in Linked Data. Despite the semantics of this property being very strict, users interpret them in a number of different and often unrelated ways. This issue will be described in more details in the following.

† The complete OWL specifications, aka OWL Full, defines a powerful knowledge representation language built on top of RDF and RDF schema. However, using OWL Full makes it possible to express statements that are (currently believed to be) "undecidable," that is, statements whose truth cannot be proved using the computational ability of any currently existing reasoning system. In fact, OWL Full (like RDF) does not enforce the strict separation of classes, properties, individuals, and datatypes into disjoint sets of resources.

In OWL Full, for instance, it is possible to state that "an eagle" represents, other than a class of individuals, an individual itself—an instance of the super-class or meta-class Eagle (Motik 2005). While this may not be an issue in RDF itself—RDF limits to *describe* resources—it is in OWL, whose purpose is to enable reasoning. OWL DL (OWL for Description Logic) is a subset of the language that, while keeping exactly the same constructs of OWL Full, enforces the aforementioned separation, matching the computational abilities of existing software with reasoning capabilities (aka *reasoners*).

Table 20.1 Common Prefixes in RDF Documents

PREFIX	URI REFERENCE
rdf:	http://www.w3.org/1999/02/22-rdf-syntax-ns#
rdfs:	http://www.w3.org/2000/01/rdf-schema#
xsd:	http://www.w3.org/2001/XMLSchema#
owl:	http://www.w3.org/2002/07/owl#

RDF classes are similar to those defined in object-oriented programming (OOP) languages. RDF specifications explicitly state that the most important difference with the latter is that while OOP languages provide prescriptions, RDF is limited to descriptions. In other words, while instance of classes, referred to as *objects* in OOP, must strictly comply with the definition of their class—for example, they cannot have attributes other than the one defined in a class—this is not necessarily true for RDF, which allows describing classes without imposing further constraints on the individual members (called extension) of the class. It is perfectly valid for an individual to be an instance of a class and present properties that are unrelated to that class or even to any other class. Contrary to attributes of OOP classes, properties in RDF are not defined in a containing class but independently from any class. Using terms from the RDF schema vocabulary it is possible, but not strictly necessary, to relate a defined property to one or more given classes.

According to the RDF schema vocabulary, resources may be classes, properties, individuals, or datatypes. The RDF predicate stating this condition is `rdf:type`. Note that `rdf:type` is itself a property that belongs to the RDF core namespace, whose prefix is usually `rdf:`. `rdf:type` has nothing special compared with other properties, including those that any RDF user can define, except its well-known meaning. For instance, the following Turtle excerpt:

```
@prefix rdf: <http://www.w3.org/1999/02/22-rdf-syntax-ns#> .
@prefix rdfs: < http://www.w3.org/2000/01/rdf-schema# > .
@prefix ex: <http://example.org/resources#> .

ex:C1 rdf:type rdfs:Class .

ex:C2 rdf:type rdfs:Class ;
  rdfs:subClassOf ex:C1 .
```

```
ex:C3 rdf:type rdfs:Class ;
  rdfs:subClassOf ex:C2 ;
  rdfs:subClassOf ex:C1 .

ex:i1 rdf:type ex:C3 ;
  rdf:type ex:C2 ;
  rdf:type ex:C1 .
```

states that `ex:C1`, `ex:C2`, and `ex:C3` are classes, that `ex:C2` is one of `ex:C1`'s subclass, and that `ex:C3` is a subclass of `ex:C2` and `ex:C1`. Finally, it also states that `ex:i1` is an instance of `ex:C3`, `ex:C2`, and `ex:C1`. Some important considerations are as follows.

The predicate `rdfs:subClassOf` in the previous statements asserts that any instance of `ex:C3` is also an instance of `ex:C2` and that any instance of `ex:C2` is also an instance of `ex:C1`.

In the RDF schema vocabulary, rdfs:subClassOf is defined as a transitive property; therefore, it is implicit that any instance of `ex:C3` (say `ex:i4`) is also an instance of `ex:C2` and `ex:C1`. Similarly it is implicit that `ex:i1`, being an instance of `ex:C3`, is also an instance of `ex:C2` and `ex:C1`. Finally, it is implicit that `ex:C2` and `ex:C3`, being subclasses, are classes. The previous RDF description is quite redundant as it contains many *materialized inferences.** It is usually summarized as follows:

```
@prefix rdf: <http://www.w3.org/1999/02/22-rdf-syntax-ns#> .
@prefix rdfs: < http://www.w3.org/2000/01/rdf-schema# > .
@prefix ex: <http://example.org/resources#> .

ex:C1 rdf:type rdfs:Class .
ex:C2 rdfs:subClassOf ex:C1 .
ex:C3 rdfs:subClassOf ex:C2 .
ex:i1 rdf:type ex:C3 .
```

These statements are not prescriptive but descriptive; the "open world" assumption holds. It could also be entirely the case that another RDF description exists, somewhere, stating that:

```
@prefix rdf: <http://www.w3.org/1999/02/22-rdf-syntax-ns#> .
@prefix rdfs: <http://www.w3.org/2000/01/rdf-schema#> .
@prefix ex: <http://example.org/resources#> .
```

* Inferences are often explicitly materialized in RDF descriptions for the sake of applications that totally lack or have limited reasoning capabilities. But even when applications do have reasoning capabilities, inferences are usually materialized, because this speeds up computations.

```
ex:C4 rdfs:subClassOf ex:C3 .
ex:i1 rdf:type ex:C4 .
```

That is, `ex:i1` is also an instance of another class in the hierarchy, say `ex:C4`, but this was not explicitly asserted in the original description.

Multiple inheritance is allowed: asserting, for instance,

```
@prefix rdf: <http://www.w3.org/1999/02/22-rdf-syntax-ns#> .
@prefix rdfs: <http://www.w3.org/2000/01/rdf-schema#> .
@prefix ex: <http://example.org/resources#> .

ex:K1 rdf:type rdfs:Class .
ex:i1 rdf:type ex:K1 .
```

which states that `ex:i1` is also an instance of class `ex:K1` (which, to the best of the knowledge this description provides, is not in the C1...C4 class hierarchy).

OWL provides richer constructs for class definitions (called axioms in OWL); first, it introduces the term `owl:Class`, which in OWL DL denotes the subset of those `rdfs:Class` that are not instances of other classes. (In OWL Full this separation is not necessary, and in fact `owl:Class` and `rdfs:Class` have the same meaning.) Second, it provides terms to state that a class is an intersection (`owl:intersectionOf`), union (`owl:unionOf`), or complement (`owl:complementOf`) of other classes or is even defined by an explicit enumeration of all its instances (`owl:oneOf`)—only for classes whose extension is enumerable. Third, in addition to `rdfs:subClassOf`, OWL provides the terms `owl:equivalentClass` stating that two classes have exactly the same class extension* and `owl:disjointWith`, to state that two class extensions have no members in common.

In RDF, properties are built-in types defined in the RDF core namespace (`rdf:`); however, they are related to classes using two RDF schema predicates: `rdfs:range` and `rdfs:domain`. The former states that all the values of a property are instances of a specified class. Similarly, the latter specified that the property could be applied only to resources that are instances of a provided class. The following Turtle excerpt:

* This OWL property, however, does not state that two classes are the same (class equality), just that instances of the one class are also instances of the other and vice versa. Stating class equality is possible using the `owl:sameAs` predicate, which is defined only for individuals. As a result, class equality is possible in OWL Full only.

```
@prefix rdf: <http://www.w3.org/1999/02/22-rdf-syntax-ns#> .
@prefix rdfs: <http://www.w3.org/2000/01/rdf-schema#> .
@prefix ex: <http://example.org/resources#> .

ex:p1 rdf:type rdf:Property ;
  rdfs:domain ex:C1 ;
  rdfs:range ex:K2 .
```

tells that `ex:p1` is a `rdf:Property` whose domain is `ex:C1` class extension and range is `ex:K2` class extension. Other than instances, properties may also have literals as range. These properties are called Datatype Properties in OWL, as opposed to the former, which are called Object Properties. The following excerpt uses the term `rdfs:Datatype` to define a property whose domain is unknown and whose range is any literal of XML schema defined type integer:

```
@prefix rdf: <http://www.w3.org/1999/02/22-rdf-syntax-ns#> .
@prefix rdfs: <http://www.w3.org/2000/01/rdf-schema#> .
@prefix xsd: <http://www.w3.org/2001/XMLSchema#> .
@prefix ex: <http://example.org/resources#> .

ex:p1 rdf:type rdf:Property ;
  rdfs:range xsd:integer .
xsd:integer rdf:type rdfs:Datatype.
```

As this example illustrates, it is perfectly legal in RDF to describe a property omitting constraints on its domain (or even on its range). The only necessary statement to define a property is the one containing the `rdf:type` predicate, saying that the resource is actually a property. On the contrary, it is possible to specify *more than one* domain and range statement. However, care must be taken in doing this, as the semantics of `rdfs:range` and `rdfs:domain` mandate that all specified class extensions are intended to be inclusive; that is, they must hold at the same time. Therefore, the following excerpt:

```
@prefix rdf: <http://www.w3.org/1999/02/22-rdf-syntax-ns#> .
@prefix rdfs: <http://www.w3.org/2000/01/rdf-schema#> .
@prefix ex: <http://example.org/resources#> .

ex:p1 rdf:type rdf:Property .
ex:p1 rdfs:range ex:C1 .
ex:p1 rdfs:range ex:K2 .
```

specifies that any value of property `ex:p1` is an instance of both `ex:K2` and `ex:C1` (this is similar to specifying more than one `rdfs:domain` for a property).

OWL allows for the definition of global cardinality constraints on properties by means of owl:FunctionalProperty (the subject of a property statement uniquely determines the object) and owl:InverseFunctionalProperty (the object of a property statement uniquely determines the subject). As an example, foaf:email is one such a property. In fact, an email (address) uniquely identifies its owner. In the following statements:

```
@prefix rdf: <http://www.w3.org/1999/02/22-rdf-syntax-ns#> .
@prefix owl: <http://www.w3.org/2000/01/rdf-schema#> .
@prefix foaf: <http://xmlns.com/foaf/0.1/> .

foaf:email rdf:type owl:InverseFunctionalProperty .
_:bn1 foaf:email <mailto:foo@example.org> .
```

the blank node _ :bn1 uniquely identifies the owner of the email address foo@example.org.*

In practice, defining the range and the domain of a property and imposing global cardinality constraints often prevent the possible usage—and reusage—of the property in contexts other than those for which it has been originally defined and adopted.

To address this issue, OWL allows for the definition of property restrictions that, rather than limiting the range of a property, identifies a class extension of individuals for which the property value meets some conditions. Seen from as an object-oriented design, this is equivalent to defining a class through its attributes, stating that for a given class, the values of each attribute are restricted to a given type. But this limitation holds only locally, that is, for the instances of that class only; globally the property may assume other values (providing they are in its range). Restrictions are of two kinds: value constraints (owl:allValuesFrom, owl:someValuesFrom, owl:hasValue); and cardinality constraints (owl:minCardinality, owl:maxCardinality, owl:cardinality). The following example defines a class ex:R1, whose instances all have the property ex:p1 constrained to assume no less and no more than three values (owl:cardinality); of these values, at least one (owl:someValuesFrom) should be of type xsd:string:

* The previous statement presented constitutes a "RDF molecule" (Ding et al. 2005), as, even if it contains a blank node, the blank node clearly identifies an entity (i.e., the owner of the email address foo@example.org).

```
@prefix rdf: <http://www.w3.org/1999/02/22-rdf-syntax-ns#> .
@prefix owl: <http://www.w3.org/2002/07/owl#> .
@prefix xsd: <http://www.w3.org/2001/XMLSchema#> .
@prefix ex: <http://example.org/resources#> .

ex:R1 rdf:type owl:Restriction ;
  owl:onProperty ex:p1 ;
  owl:someValuesFrom xsd:string ;
  owl:cardinality "3"^^xsd:nonNegativeInteger .
```

As previously stated, properties are treated as full "first-category" resources in RDF and have a global scope. It is even possible to define a hierarchy for them (as with classes). All subproperties share the same domain and range of their parent properties. In following example a *subproperty* is defined:

```
@prefix rdf: <http://www.w3.org/1999/02/22-rdf-syntax-ns#>.
@prefix rdfs: <http://www.w3.org/2000/01/rdf-schema#>.
@prefix ex: <http://example.org/resources#>.

ex:p1 rdf:type rdf:Property.
ex:p2 rdfs:subPropertyOf ex:p1.
```

`rdfs:subPropertyOf` is transitive. This fact itself can be expressed in OWL using the following statements:

```
@prefix rdf: <http://www.w3.org/1999/02/22-rdf-syntax-ns#>.
@prefix rdfs: <http://www.w3.org/2000/01/rdf-schema#>.
@prefix owl: <http://www.w3.org/2002/07/owl#>.

rdfs:subPropertyOf rdf:type owl:TransitiveProperty.
```

OWL also makes it possible to state that two properties are equivalent; they share the same extension (i.e., the pairs: subject–object) of all statements where the property appears as a predicate using the term `owl:equivalentProperty`.* It is also possible to state that one of the properties is the inverse of the other through the OWL property `owl:inverseOf`, which is symmetrical; this fact is expressed in OWL using the statements:

```
@prefix rdf: <http://www.w3.org/1999/02/22-rdf-syntax-ns#>.
@prefix rdfs: <http://www.w3.org/2000/01/rdf-schema#>.
@prefix owl: <http://www.w3.org/2002/07/owl#>.

owl:inverseOf rdf:type owl:SimmetricProperty.
```

* The same caveats described for `owl:equivalentClass` apply.

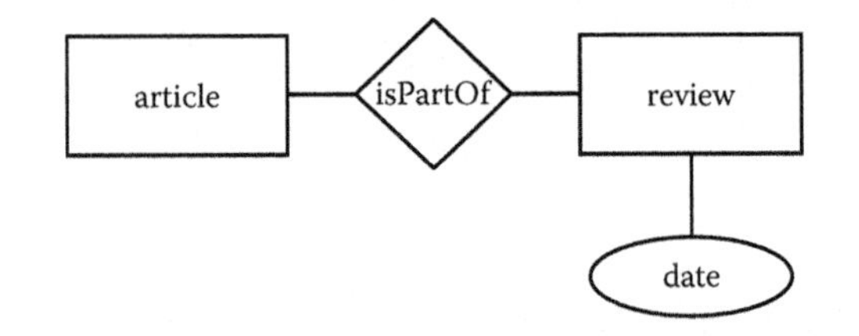

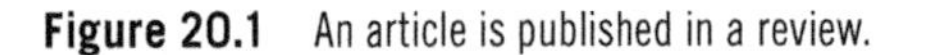

Figure 20.1 An article is published in a review.

RDF schema provides two useful properties for annotating resources: `rdfs:label` and `rdfs:comment` are used, respectively, to provide human-readable name and description for resources.

Reification, Quads, and Named Graphs

Despite its expressivity, RDF statements are all based on simple binary relationships between subject and object. RDF is thus unsuitable to model *n*-ary relationship. Figure 20.1 illustrates a possible entity–relationship diagram (ERD; a widely used representation in relational database design), which may model the following relationship: an article is part of a review published on a given date.

To deal with this statement, RDF needs to materialize the resource identifying the review using either a URIRef or a blank node and to assign the property `date` to this node (instead of assigning this property to the relationship that connects the article to the review). This is illustrated in the following example (Figure 20.2), using terms from the Dublin Core Metadata Initiative vocabulary (prefix `dcterms:`; see Table 20.2)—one of the first vocabularies accepted by the Semantic Web community and now a key collection of terms in linked data.

```
@prefix dcterms: <http://purl.org/dc/terms/>.
@prefix ex: <http://example.org/resources#>.

ex:aPaper dcterms:title "An Example Paper"^^xsd:string;
  dcterms:isPartOf _:aReview.
_:aReview dcterms:title "Examples"^^xsd:string ;
  dcterms:identifier "urn:issn:nnnn -mmmm"
  dcterms:date "2008-01-20"^^xsd:date.
```

In this case, modeling a node for the review hosting the article and giving it the attribute `date` may appear to be the most reasonable

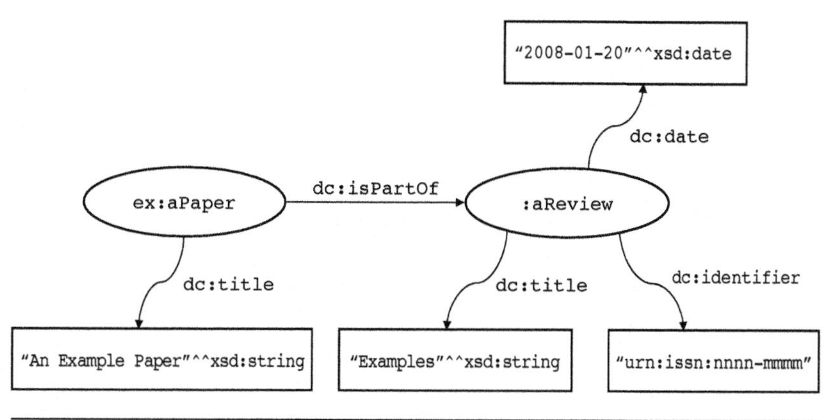

Figure 20.2 The article titled "An Example Paper" is part of the review "Examples" published January 20, 2008.

Table 20.2 Other Common Prefixes in RDF Documents

COMMON PREFIX	URI REFERENCE	DESCRIPTION
dc: dcterms:	http://purl.org/dc/elements/1.1/ http://purl.org/dc/terms/	The *Dublin Core vocabulary* provides broad and generic terms for describing a wide range of resources.
cc:	http://creativecommons.org/ns#	The *Creative Commons vocabulary* contains terms that state copyrights and describe licenses.
void:	http://rdfs.org/ns/void#	The *Vocabulary of Interlinked Datasets* provides terms that describe datasets, how they are exposed, and their relationships with other datasets.
skos:	http://www.w3.org/2004/02/skos/core#	The *Simple Knowledge Organization Systems vocabulary* provides terms that describe taxonomies (i.e., conceptual hierarchies).
foaf:	http://xmlns.com/foaf/0.1/	The *Friend-of-a-Friend vocabulary* provides terms for describing people, activities, organizations, and their relationships.

choice. However, there are other cases in which the lack of support for *n*-ary relationships in RDF is evident,* such as the assertion that an employee works for a company since January 20, 2008. A corresponding ERD is depicted in Figure 20.3. This figure depicts a fact with temporal validity. Unfortunately, RDF does not provide support for temporal aspects (despite several ongoing attempts to standardize

* A more colorful example is reported in Brickley (2011).

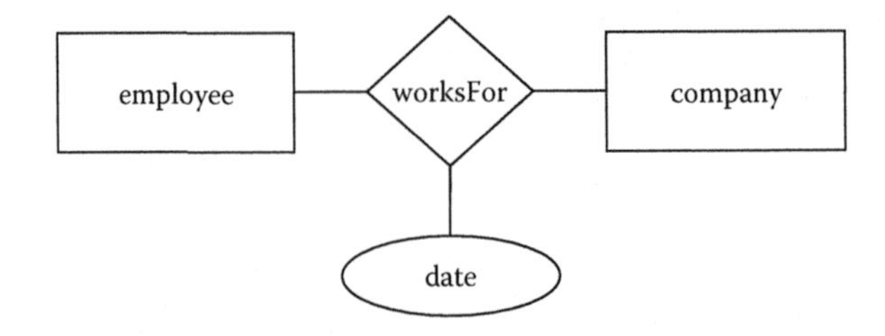

Figure 20.3 An employee works for a company since a hiring date.

schemas accounting for temporal aspects), and there is no way to attach a property (e.g., `dcterms:date`) to an arc in the RDF graph.* So this assertion is simply not expressible in RDF.

More in general, RDF itself does not provide any formal support for handling contextual aspects, despite the fact that there exists a well-known reification vocabulary allowing the description of an RDF statement (which can be thought of as a resource as well). For instance, the following reification quad:

```
@prefix rdf: <http://www.w3.org/1999/02/22-rdf-syntax-ns#>.
@prefix foaf: <http://xmlns.com/foaf/0.1/>.
@prefix ex: <http://example.org/resources#>.

ex:aTriple rdf:type rdf:Statement ;
  rdf:subject ex:person1 ;
  rdf:predicate foaf:knows ;
  rdf:object ex:person2.
```

describes a statement whose subject is `ex:person1`, object is `ex:person2`, and predicate is `foaf:knows`.† However, the reification describes *one possible* statement, which *might be* asserted, but not *one specific* triple in a *given* RDF document. In fact, nothing prevents one from thinking that the URIRef `ex:aTriple` in the previous example identifies a specific RDF triple stored somewhere. And, even if this were possible, there would be no means to identify the predicate `foaf:knows` inside that triple. Together with other less popular

* This need emerged in the context of the World Wide Web Consortium (W3C) Provenance Working Group (http://www.w3.org/2011/prov/) as one possible requirement behind a standardized provenance model.

† Created in 2000, the Friend-of-a-Friend (FOAF) vocabulary, together with the Dublin Core Metadata Initiative, is one of the most known and used schema in the Semantic Web and linked data.

features (e.g., RDF containers), Tim Berners-Lee (2010) suggested deprecating reification in a recent note.

While reification does not provide a useful "hook" to bind RDF triples to a specific context, the RDF community with time has elaborated several different alternatives that might provide this preferred feature. For example, N-Quads (Cyganiak, Harth, and Hogan 2008) seem to have emerged as a viable, widely implemented solution. Basically, an N-Quad is a simple extension of an RDF triple with a fourth element, a URIRef identifying the source of the RDF statement (Dumbill 2003). This choice has the advantage of keeping separate triples that have the same subject, predicate, and object but come from different data sets. In addition, it has recently been proposed that quad might provide support for *anchored* properties, that is, properties whose validity is limited to the context where they appear (Hayes 2011).

Named graphs extend the idea behind quads but modify the syntax in favor of a more clearly defined semantic for the source-identifier element. The original idea behind named graphs comes from the observation that an RDF graph is intentionally *floating*, that is, not formally bound to any identifier or location where one of its representation can be found. If such a facility existed, one would be able to use that URIRef to contextualize the assertions contained within that RDF graph. Named graphs provides this capability: instead of having an identifier as an additional fourth element in each RDF triple, as in quads, named graphs provide an identifier for the whole graph.

A named graph is not an RDF graph; rather, it is a pair (name–graph). In this pair, the *name* is a URIRef, whereas the *graph* is one of the many possible equivalent RDF graphs.* Two RDF graphs that differ only in their blank nodes are considered equivalent RDF graphs; however, they are part of two different named graphs. Christian Bizer, one of the proponents of named graphs, observes that the relationship between a named graph and a corresponding RDF graph is the same occurring between an instance and its class: there might be infinite instances of one single class; similarly there might be infinite,

* In fact, given an original graph, one may create an equivalent graph that differs from the original for its blank nodes only. Therefore, there may be an infinite number of equivalent graphs.

equivalent, individual named graphs corresponding to the same RDF graph but actually differing from each other in their name and, possibly, in their blank nodes (Carroll et al. 2005).

Two concrete syntaxes for named graphs are TriX and TriG. Respectively, they extend RDF/XML and Turtle.

The name of a named graph is a URIRef, which may be freely chosen like any other resource. For legacy RDF graphs published on the web, however, it is convenient to use the Uniform Resource Locator (URL) where the graph (or the value of the first `xml:base` attribute in the corresponding RDF/XML document) is retrieved. The following is an example of a graph named after its location <http://example.org/graph1>, expressed in TriG syntax:

```
@prefix dcterms: <http://purl.org/dc/terms/>.
@prefix xsd: <http://www.w3.org/2001/XMLSchema#>.
@prefix ex: <http://example.org/>.
@prefix res: <http://example.org/resources#>.

ex:graph1 {
_:paper1 a res:Paper ;
   dcterms:title "The First Example Paper"^^xsd:string ;
   dcterms:isPartOf _:aReview.
_:aReview a res:Review ;
   dcterms:title "Examples"^^xsd:string ;
   dcterms:identifier "urn:issn:nnnn -mmmm" ;
ex:graph1
   dcterms:date "2010-10-12"^^xsd:date.
}
```

Inside this named graph the RDF triple

```
ex:graph1 dcterms:date "2010-10-12"^^xsd:date.
```

describes the publication date of the graph itself. In fact, named graphs are RDF resources that can be described in RDF triples as any other resource, and the description can even be included in the graph itself.

RDF and XML

To understand why RDF has been introduced and what its benefits are compared with pure markup languages, it is useful to present a comparison with Extensible Markup Language (XML). RDF and XML share some important features and present similarities.

Both use unique identifiers and namespaces to keep the identifiers separated and to prevent possible confusion in interpretation. In XML elements are identified by fully qualified name (FQN), in form of

```
namespace:localname
```

This syntax is also explicitly chosen for RDF where URIRefs are represented as FQN. Both the Uniform Resource Name (URN) and the URL schema can be used in FQN. Using the URL schema allows applications to dereference the URIRef to obtain additional information (e.g., retrieving the XML schema behind an XML document or getting a web page containing the human-readable description of an RDF resource). When the URN schema is used, an external resolver may be provided to obtain a similar effect, but this resolver introduces a new, redundant software layer that can be avoided by using URL. Due to this reason, Tim Berners-Lee has recommended using a Hypertext Transfer Protocol (HTTP) URI (i.e., URL) as the best practice for linked data.

Both are open and extendible standards. Anyone can create an XML document as well as an RDF description using no more than a simple text editor, and anyone can create an XML schema or an RDF schema. XML schemas are, however, closed and prescriptive; that is, they cannot be validated if unknown elements are found in the corresponding schema instance. RDF schemas are descriptive, more "tolerant," and based on the open-world assumption that new properties not originally enumerated in the schema can be freely added as long as they do not contradict the schema itself. XML schema accomplish this function with the use of the `xsd:anyURI` special element (which allows any element to be nested within a given XML element in the corresponding schema instance), an exceptional case that should be explicitly declared. In RDF this behavior is the default.

Both XML and RDF allow linkage of remote objects, but there is a slight difference. While in RDF linking is a native feature (the RDF predicate implicitly links the subject with the object and each of them may reside in any part of the Network), in XML a similar capability was not part of the core language but was added subsequent to its creation (DeRose, Maler, and Orchard 2001). Meanwhile, different technologies have adopted different linkage mechanisms. For instance, in XDM, shared XDMs use a linkage technique inside the `<service-list>` element. The `<service-list>` is actually a container for a number of elements in the form

```
<entry uri = "http://example.org/resource">
```

with the attribute `uri` used to specify the URI where the service can be found.

RDF does not define datatypes; rather, it reuses some, but not all, XML schema datatypes. A list of the datatypes acceptable in RDF is shown in Klyne, Carroll, and McBride (2012).

Semantically, however, data in RDF are on a higher level of abstraction than XML data. This is evident in a number of marked differences between the two languages. For example, RDF is resource oriented, while XML is representation oriented. XML would not exist without concrete text serialization (i.e., its representation). RDF is based on a more abstract model (the RDF graph) that can materialize into a representation with a concrete syntax, RDF/XML, or Turtle, for example, but might even exist without having a defined a priori representation. The choice to avoid datatype definition in favor of referencing existing ones (those defined in XML schema) is a step in this direction.

The only composition semantics in XML is the expression of containment or composition that is incorporated by placing elements within elements according to a strict linear hierarchy. XML does not allow for the expression of different nonhierarchical relationships, such as equivalence.

In RDF, properties are *first-class objects* with their own hierarchy and attributes. For example, in RDF it is possible to express that a property is inverse functional or that two properties are equivalent. The concept of independently defined relationship is largely missing in XML where attributes and subelements are always hierarchically dependent on a father element.

Not allowing the definition of RDF properties to be subordinate to those of another element has permitted the reuse of RDF properties in a number of different schemas. This contributes to increased consolidation of the commonly understood core RDF dictionaries.

With the emergence of linked data, many organizations have started porting XML data into RDF. This translation is often called *lifting*, while the opposite process is called *lowering*. XQuery and XSLT along with SPARQL are useful tools for lifting, as originally suggested in GRDDL (Connolly 2007) and SAWSDL (Farrell and Lausen 2007) specifications. For what concerns lowering, XSPARQL, merging the advantages of XQuery with the ones from SPARQL, has been indicated as a viable solution. XSPARQL has been submitted to W3C along with a number of use cases that, other than covering lifting and lowering, are also intended to enhance RDF to RDF mapping capabilities not originally present in the SPARQL CONSTRUCT.

21
RDF Query Language: SPARQL

Whereas Resource Description Framework (RDF) statements can be thought of as a graph or collection of graphs, querying RDF means to be able to detect portions of those graphs matching given patterns. RDF has its own query language, SPARQL, which provides a rich set of capabilities for this purpose.

Triple Patterns and Query

SPARQL patterns are expressed using a syntax similar to SQL but which is based on the concept of triple patterns. Triple patterns are similar to RDF triples except that each subject, object, and predicate in a triple pattern may be a variable.

A given pattern matches a portion of a graph, or subgraph, if and only if a graph equivalent to that subgraph may be obtained by substituting the variables that appear in the pattern with corresponding terms from the subgraph. A query can return zero, one or more solutions that associate actual terms with the selected variables. Each existing solution is intended as a possible match of the pattern of a given subgraph, which is present in the input data set.

For instance, given the following data set:

```
@prefix dcterms: <http://purl.org/dc/terms/> .
@prefix ex: <http://example.org/resources#> .
@prefix xsd: <http://www.w3.org/2001/XMLSchema#> .

_:paper1 a ex:Paper ;
   dcterms:title "The First Example Paper"^^xsd:string ;
   dcterms:isPartOf _:aReview .
_:paper2 a ex:Paper ;
   dcterms:title "The Second Example Paper"^^xsd:string ;
   dcterms:isPartOf _:aReview .
```

```
_:paper3 a ex:Paper ;
   dcterms:title "The Third Example Paper"^^xsd:string .
_:aReview a ex:Review ;
   dcterms:title "Examples"^^xsd:string ;
   dcterms:identifier "urn:issn:nnnn -mmmm" ;
   dcterms:date "2004-12-06"^^xsd:date .
```

the following SPARQL query

```
PREFIX dcterms: <http://purl.org/dc/terms/>
PREFIX ex: <http://example.org/resources#>
SELECT ?paperTitle ?reviewTitle
WHERE {
  ?a a  ex:Paper ;
     dcterms:title ?paperTitle ;
     dcterms:isPartOf ?review .
  ?review dcterms:title ?reviewTitle .
}
```

uses the variables `?paperTitle`, `?reviewTitle`, `?a`, and `?review`* to define a graph pattern intended to match only articles that appear in a review. Once done, the titles of the articles and reviews are returned. When executed on the previous data set, the query produces a solution set made of two entries:

paperTitle	**reviewTitle**
"The First Example Paper" ^^http://www.w3.org/2001/XMLSchema#string	"Examples" ^^http://www.w3.org/2001/XMLSchema#string
"The Second Example Paper" ^^http://www.w3.org/2001/XMLSchema#string	"Examples" ^^http://www.w3.org/2001/XMLSchema#string

In fact, only two of the three articles in the data set are stated to be part of a review. (Since nothing is said about the third articles, the SPARQL processor does not find any match for the specified graph pattern.) Note also that, as specified in the pattern, only resources declared to be articles are considered, while reviews are not.

In the previous example, two extra variables, `?a` and `?review`, are used in the graph pattern but are not selected. Actually, they could

* The syntax used for variables is a question mark (?) followed by variable name; due to historical reasons, alternatively, it is possible to use the dollar symbol ($) instead of the question mark, a syntax similar to the one used in XQuery.

be replaced by a blank node. SPARQL allows alternative syntaxes for blank nodes; for example, ?a might be replaced by

```
_:_
_:a
[]
[a ex:Paper]
```

In particular, the latter is a shortcut that may serve as a subject or object of another statement; thus, the previous query might be expressed in a more compact form, as in the following query. This query also introduces the built-in function regex(), matching regular expressions, and the keyword FILTER, used to define conditions that may restrict the solution set. Only those solutions for which the filter expression is evaluated to true are returned.

```
PREFIX dcterms: <http://purl.org/dc/terms/>
PREFIX ex: <http://example.org/resources#>
PREFIX xsd: <http://www.w3.org/2001/XMLSchema#>

SELECT ?paperTitle ?reviewTitle
WHERE {
  [a ex:Paper] dcterms:title ?paperTitle ;
    dcterms:isPartOf [dcterms:title ?reviewTitle].
    FILTER (regex(?paperTitle, "Sec"^^xsd:string))
}
```

When executed this query returns only one row:

paperTitle	reviewTitle
"The Second Example Paper" ^^http://www.w3.org/2001/ XMLSchema#string	"Examples" ^^http://www.w3.org/2001/ XMLSchema#string

Other than in tabular format, solutions may be returned in the form of an RDF graph using the CONSTRUCT query form. Inside a CONSTRUCT block it is possible to specify a template that the SPARQL processor will use to produce the output graph. This query form is particularly useful to manipulate RDF graphs. For example, the following query:

```
PREFIX dcterms: <http://purl.org/dc/terms/>
PREFIX ex: <http://example.org/resources#>
CONSTRUCT {
  ?a dcterms:title ?t ;
    dcterms:date ?d.
```

```
}
WHERE {
  ?a a  ex:Paper ;
     dcterms:title ?t ;
     dcterms:isPartOf [dcterms:date ?d].
```

extracts from the previous data set titles and date of publication of articles that are part of the review. The date of publication is taken from the date of publication of the review. The result is the following output graph:

```
@prefix dcterms: <http://purl.org/dc/terms/>.
@prefix ex: <http://example.org/resources#>.
@prefix xsd: <http://www.w3.org/2001/XMLSchema#>.

[] dcterms:date "2004-12-06"^^xsd:date ;
   dcterms:title "The Second Example Paper"^^xsd:string.

[] dcterms:date "2004-12-06"^^xsd:date ;
   dcterms:title "The First Example Paper"^^xsd:string.
```

Graph Patterns

The graph patterns appearing in the previous queries within the WHERE expression (including any optional FILTER clause whose scope is limited to the graph pattern where it appears) are called basic graph patterns, but there might be more complex patterns. For instance, group graph patterns allow for the grouping of a set of statements into separate graph patterns, which may also be nested. The following query is equivalent to the previous one, but uses two basic graph patterns (one nested into another):

```
PREFIX dcterms: <http://purl.org/dc/terms/>
PREFIX ex: <http://example.org/resources#>
SELECT ?paperTitle ?reviewTitle
WHERE {
  ?a a  ex:Paper ;
     dcterms:title ?paperTitle ;
     dcterms:isPartOf ?review.
   {
     ?review dcterms:title ?reviewTitle.
   }
}
```

The convenience of arranging statements into different basic graph patterns—instead of having a single basic graph pattern—is evident when considering other SPARQL features, such as unions, optional graph patterns, and negations. Unions are useful to match alternative patterns. For example, the following query returns titles of resources that may be either articles or reviews:

```
PREFIX dcterms: <http://purl.org/dc/terms/>
PREFIX ex: <http://example.org/resources#>
SELECT ?title
WHERE {
{
   [dcterms:title ?title]   a  ex:Paper.
} UNION {
   [dcterms:title ?title]   a  ex:Review.
}
}
```

Optional graph patterns allow the declaration that some statements may optionally be present, which is a particularly useful feature when working with remote data sets containing partial and a priori unknown information. The following query matches all articles optionally belonging to a review and returns their titles:

```
PREFIX dcterms: <http://purl.org/dc/terms/>
PREFIX ex: <http://example.org/resources#>
SELECT ?paperTitle ?reviewTitle
WHERE {
  ?a a  ex:Paper ;
     dcterms:title ?paperTitle.
     OPTIONAL {
        ?a dcterms:isPartOf [dcterms:title ?reviewTitle].
     }
}
```

The result, reported as follows, clearly shows that the ?reviewTitle variable is not bound to a particular value in one of the returned solution. A variable not bound to a particular value is called an *unbound variable.*

paperTitle	reviewTitle
"The First Example Paper" ^^http://www.w3.org/2001/ XMLSchema#string	"Examples" ^^http://www.w3.org/2001/ XMLSchema#string
"The Second Example Paper" ^^http://www.w3.org/2001/ XMLSchema#string	"Examples" ^^http://www.w3.org/2001/ XMLSchema#string
"The Third Example Paper" ^^http://www.w3.org/2001/ XMLSchema#string	–

SPARQL provides a function, bound(), which can be used to test whether a variable is bound.

Other than in optional graph patterns, unbound variables may typically appear in negations, which are another possible way of combining basic graph patterns. Negations allow the definition of graph patterns that must not be matched by the query. The following query selects only articles that have not been declared to be part of a review. This query puts an unwanted graph pattern inside an OPTIONAL block and uses the function bound() and the boolean operator NOT, expressed as the bang symbol ("!"), to filter out solutions that fit the optional graph pattern:

```
PREFIX dcterms: <http://purl.org/dc/terms/>
PREFIX ex: <http://example.org/resources#>
SELECT ?paperTitle
WHERE {
  ?a a  ex:Paper ;
     dcterms:title ?paperTitle.
  OPTIONAL {
     ?a dcterms:isPartOf ?review.
  }
  FILTER (!bound(?review))
}
```

As a shorter equivalent form, the following query encloses the unwanted match within a NOT EXISTS filter expression:

```
PREFIX dcterms: <http://purl.org/dc/terms/>
PREFIX ex: <http://example.org/resources#>
SELECT ?paperTitle
WHERE {
  ?a a  ex:Paper ;
    dcterms:title ?paperTitle.
  FILTER NOT EXISTS {
     ?a dcterms:isPartOf ?review.
  }
}
```

An alternative way to obtain a similar effect is to use the keyword MINUS. However, this operator works differently. Instead of discarding solutions for which a specified graph pattern is matched, MINUS evaluates both the graph pattern to be matched and the one not to be matched and then computes the logical difference between the two solution sets. Using MINUS the previous query is rewritten as

```
PREFIX dcterms: <http://purl.org/dc/terms/>
PREFIX ex: <http://example.org/resources#>
SELECT ?paperTitle ?reviewTitle
WHERE {
  ?a a  ex:Paper ;
     dcterms:title ?paperTitle.
  MINUS {
     ?a dcterms:isPartOf ?review.
  }
}
```

SPARQL inherits from SQL many features. For instance, an ORDER BY clause may be added to specify the order of the solution sequence. The DISTINCT modifier is used to ensure that solutions in a query are unique. As in SQL, queries may contain aggregates (e.g., COUNT, SUM, MIN, MAX), which apply expression over groups of solutions. Aggregates use the GROUP BY clause, which partitions the solution into groups. Then the aggregate values are computed for each group. The keyword HAVING is used to filter out groups of solutions for which the expression defined in the HAVING clause is false.

Querying from Multiple Graphs

In the usual practice, the owner of a given data set publishes a SPARQL endpoint, for example, a Uniform Resource Locator (URL) where a user interface of the SPARQL processor can be found that is able to trigger and execute a query over the data set. In general, however, SPARQL allows the programmer to specify the composition of the data set on which the query will be executed, in terms of graphs. Each data set in fact contains at least one default unnamed graph and may contain additional named graphs. The graphs are retrieved from their sources, which are identified by a URL. Each URL points to a representation of the graph. To instruct the SPARQL processor to fetch a representation available at the specified URL, the FROM keyword is used. When multiple FROM clauses are used, the obtained graphs are merged into one single default graph. (Blank nodes that share the same local identifiers are renamed to keep them separate.) The query is executed on the latter.

To support named graphs, SPARQL introduces the FROM NAMED keyword. If one or more FROM NAMED clauses are used, then each named graph is kept separate from the other(s) and from the default graph (which, by default, is always present in the data set). When this happens, the GRAPH keyword is used to make active one named graph in the graph pattern to be matched. The GRAPH keyword is followed either by the name of the graph (typically the URL at which its representation can be retrieved) or by a variable.* The first usage restricts the graph pattern matching to a given named graph; the latter has several applications, including the possibility to restrict pattern matching to only graph names that satisfy a given triple pattern.

The following query uses three named graphs and one default graph. The three named graphs, located at <http://example.org/resources#graph-n>, contain statements about articles and the reviews in which they are published:

```
#First named graph, located at http://example.org/
resources#graph-1
@prefix dcterms: <http://purl.org/dc/terms/>.
@prefix ex: <http://example.org/resources#>.
@prefix xsd: <http://www.w3.org/2001/XMLSchema#>.

ex:graph-1 {
_:paper1 a ex:Paper ;
    dcterms:title "The First Example Paper"^^xsd:string ;
    dcterms:isPartOf [
    a ex:Review ;
      dcterms:title "Examples 1"^^xsd:string] .
}

#Second named graph, located at http://example.org/
resources#graph-2
@prefix dcterms: <http://purl.org/dc/terms/>.
@prefix ex: <http://example.org/resources#>.
@prefix xsd: <http://www.w3.org/2001/XMLSchema#>.

ex:graph-2 {
_:paper2 a ex:Paper ;
    dcterms:title "The Second Example Paper"^^xsd:string ;
    dcterms:isPartOf [
    a ex:Review ;
      dcterms:title "Examples 2"^^xsd:string] .
}
```

* Graph names, in fact, are treated as any other RDF terms and as such may be values of variables and may be returned in the solution set.

```
#Third named graph, located at http://example.org/
resources#graph-3
@prefix dcterms: <http://purl.org/dc/terms/>.
@prefix ex: <http://example.org/resources#>.
@prefix xsd: <http://www.w3.org/2001/XMLSchema#>.

ex:graph-3 {
_:paper3 a ex:Paper ;
    dcterms:title "The Third Example Paper"^^xsd:string.
}
```

The default, unnamed graph is located at <http://example.org/resources#index> and contains triples that state the publishing date of the three named graphs:

```
ex:graph-1 dcterms:date "2009-02-12"^^xsd:date.
ex:graph-2 dcterms:date "2010-10-04"^^xsd:date.
ex:graph-3 dcterms:date "2011-05-07"^^xsd:date.
```

Given this input data set, the following query returns only solutions taken from named graphs published after the year 2009. This query uses the built-in function `year()`, which returns the year of a `xsd:date` typed literal:

```
PREFIX dcterms: <http://purl.org/dc/terms/>
PREFIX ex: <http://example.org/resources#>

SELECT ?paperTitle ?reviewTitle ?source

FROM <http://example.org/index>
FROM NAMED <http://example.org/graph-1>
FROM NAMED <http://example.org/graph-2>
FROM NAMED <http://example.org/graph-3>

WHERE {
  GRAPH ?source {
  [a ex:Paper] dcterms:title ?paperTitle ;
    dcterms:isPartOf [dcterms:title ?reviewTitle].
  }

  ?source dcterms:date ?d.
  FILTER (year(?d)>2009)
}
```

The result is

paperTitle	reviewTitle	Source
"The Second Example Paper" ^^http://www.w3.org/2001/XMLSchema#string	"Examples 2" ^^http://www.w3.org/2001/XMLSchema#string	http://example.org/resources#graph-2

22
Linking Open Data Project

The Linking Open Data community project started in January 2007 at the World Wide Web Consortium (W3C). Its original aim was to identify existing open data sets and to publish them on the web as linked Resource Description Framework (RDF) documents. The chosen data sets were all *open data,* that is, data available under open license or factual data.* Choosing the concept of open data resulted in the creation of the Creative Commons licenses, which enables authors to explicitly state conditions for other stakeholders to reuse their work, as well as the Open Data Commons Public Domain Dedication and License (PDDL), which covers factual data that by definition cannot be copyrighted. Table 22.1 summarizes the terms of these licenses.

Subsequently, the number of available data sets increased exponentially with many stakeholders including universities, libraries, open-source software communities, media and news stakeholders, commercial corporations, and even government agencies, subject to the conditions raised by the Open Government Partnership initiative.† At the end of 2011 the total amount of RDF triples in the Linked Open Data cloud was estimated to be about 31 billion.

* According to the Open Knowledge Foundation (http://okfn.org/), a piece of content or datum is open if anyone is free to use, reuse, and redistribute it—subject only, at most, to the requirement to attribute or share alike.

† The Open Government Partnership project (http://www.opengovpartnership.org/) was formally launched in 2011 and was subsequently extended with the commitment of forty-six national governments. One of the key enablers of the Open Government Partnership involves the possibilitiy for third parties to freely access and reuse open-government information for purposes other than the original intent for which the data were collected. Such information should be provided as nonproprietary, patent-free, machine-readable data.

Table 22.1 Creative Commons and Open Data Commons Licenses

LICENSE	SUMMARY
Creative Commons CC-BY-SA 3.0	CC-BY-SA 3.0 makes it possible to copy, distribute, transmit, and adapt a creative work and to make commercial use of it; it is mandatory to redistribute the adapted work under the same or similar license. It is mandatory to attribute the work to the author.
Creative Commons CC-BY 3.0	CC-BY 3.0 makes it possible to copy, distribute, transmit, and adapt a creative work and to make commercial use of it. It is mandatory to attribute the work to the author.
Creative Commons CC-Zero 1.0	CC-Zero 1.0 makes it possible to copy, modify, distribute, and perform a creative work, even for commercial purposes, without asking any permission and without attributing the work to the author.
Open Data Commons ODbL 1.0	ODbL 1.0 makes it possible to share and adapt the data contained in a database and to create and distribute a derivative work as long as it is kept open and under the same license. It is mandatory to attribute the work to the author.
Open Data Commons ODC-BY-SA	Unlike other licenses, ODC-BY-SA defines social expectations, not legal requirements. Data or the derivative works should be shared alike, attributed to the author, and kept open. Data access should not be restricted.
Open Data Commons PDDL	PDDL eliminates any restriction held by the original creator of the data. It makes it possible to share and adapt the data and to create and distribute a derivative work without asking permission nor providing any attribution. This license explicitly includes commercial use, and does not exclude any other field of endeavor.

This vast amount of linked open data has doubled its size yearly.* From the beginning, a publicly available catalogue of data sets has been maintained by the linked data community and published on the Comprehensive Knowledge Archive Network (CKAN).†

Classifying Open Data: The Five-Star Rating System

To explain the migration of electronic data to linked data, Tim Berners-Lee proposed a five-star rating classification system. In it,

* A regularly updated diagram tracing the evolution of the Linked Open Data cloud is maintained by Richard Cyganiak and Anja Jentzsch and is available for free at http://richard.cyganiak.de/2007/10/lod/.

† The Comprehensive Knowledge Archive Network, http://www.ckan.net/.

Table 22.2 Five-Star Rating

LICENSE	SUMMARY
★☆☆☆☆	Data on the web, encoded in any format
★★☆☆☆	Data on the web encoded in any machine-readable format
★★★☆☆	Data on the web in any open, standard format (e.g., CSV, XML, JSON, RDF)
★★★★☆	Data on the web in an open format addressable by URI at the finest level of granularity (RDF)
★★★★★	Linked Open Data on the web

available electronic data are classified according to format and availability for third-party users. Each additional star in the classification schema adds possibilities in terms of data management, and reuse of data classified with fewer stars is prohibited. Hausenblas (2010) discusses costs and benefits of each kind of data.

Table 22.2 summarizes the proposed classification schema. Single-star data characterize information available on the web as unstructured data with an open licence. This information can be reused but typically at the cost of manual adaptation. Examples include text documents, diagrams, pictures, and other rendered objects with unavailable sources. Two stars indicates data that, in addition to the aforementioned characteristics, are available as structured data in a proprietary format. The most representative examples of this category available worldwide include data in database tables and spreadsheets. These data can be automatically processed but at the cost of having proprietary software tools that enable their handling. Most of these tools, however, enable conversion into other formats, including nonproprietary ones (i.e., the comma-separated value [CSV] format or Extensible Markup Language [XML]). In this classification schema, data available under standard nonproprietary formats are given three stars.

Data published as RDF data sets are classified with four stars in the rating system. Compared with XML, RDF adds the ability to natively identify and address data at the finest level of granularity. Publishing in RDF, however, comes with the cost of minting URIRefs for individuals and investing in seeking out (or inventing if nonexistent) suitable vocabularies for the classes and properties found in the data set to be published. Some previously existing tools may assist the publisher in this process. To facilitate this task, the W3C RDB2RDF Working Group has recently released two different proposals for a standard language to express the mapping of relational database tables into RDF (Arenas et al. 2012; Das, Sundara, and Cyganiak 2012). The two proposals essentially differ in the degree of customizability of the mapping.

Five stars are obtained by open data that are published as an RDF data set and are linked to other (RDF) data sets. Integrating different data sets may be costly and time-consuming, but once they are done the integrated data sets serve as an ideal "giant RDF graph" and could be used to answer queries that were impossible before data sets were linked. The integration happens progressively in a "pay-as-you-go" approach. The more data sets are integrated, the higher the value found in the linked data cloud dataspaces (Cafarella, Madhavan, and Halevy 2009).

Four Linked Data Principles

Web application programming interfaces (APIs) are currently the fashionable and easy way of accessing structured data on the web. But even if web APIs are based on the common Hypertext Transfer Protocol (HTTP) and on the well-accepted Representational State Transfer (REST) paradigm, the main problem is that each data set has a provider-specific nonstandardized schema. Applications that use different data sources on the web (commonly referred to as *mashups*) have to rely on their own code to link data from one provider to data from another. Separate code is needed for each pair of exposed data sets. So even if access to different data sets on the web is granted, the lack of standardized semantics causes a compatibility problem that must be solved each time with a different solution. Furthermore, each data set remains isolated from the others and is thus prone to possible redundancies and inconsistencies.

Berners-Lee (2006) enunciated the four principles involved in the move from isolated data sets to a global data space over the web (i.e., linked data). The first principle states that URIRefs should be used as identifiers for objects. This solves the problem of heterogeneous naming schemas used in different APIs to identify objects. URIRefs provide a uniform mechanism to refer to objects of interest in the data set that are then exposed as resources. The second principle further specifies that the schema adopted for those URIRefs should be HTTP. The advantage of HTTP compared with other schemas is that HTTP is well supported, is decentralized, and does not need any external resolution mechanism, as HTTP Uniform Resource Identifiers (URIs) are already automatically resolved by the Domain Name System (DNS). Additionally, resources identified by HTTP URIRefs are accessible, and any HTTP client can dereference the URIRef and provide access to a resource, such as a web page containing information describing the identified entity.

This fact is actually stated in the third principle: provide useful information about the referenced entity in a standard format (i.e., RDF). Several techniques can be used to do this. One possibility is to include an XML document conforming to a sitemap in the web page by dereferencing the HTTP URIRefs associated with the resource. Traditionally, sitemaps are stored in the root directory of web sites and describe the site itself.* An extension of the sitemaps protocol can be used to convey information about the HTTP URI where the original dataset resides or the URI of a SPARQL endpoint that can be used to query that data set. Another possibility is to use HTTP content negotiation to preset either a web page or an RDF document based on the HTTP client's request (this technique will be explained later).

The fourth and final principle mandates that the useful information returned should include links to other related entities so that a user can discover related resources by browsing these links the way pages are browsed on the traditional web. Using the five-star rating system, it is possible to combine the third and fourth principle: provide five-star data about the referenced entity.

Cool URIs for the Semantic Web

The URIRef used to refer to an entity should be different from the one used to access its possible description. The "httpRange-14" decision (Fielding 2005) provides a clarification of this topic introducing the terms *noninformation resources* and *information resources*. Information resources are resources that convey all their essential characteristics (Jacobs and Walsh 2004) in the representations returned in response to a HTTP request. This does not hold true for noninformation resources.

* Typically, sitemaps are crawled by search engines to retrieve hints about the pages contained in the site.

Data publishers should use at least two different URIRefs: one to refer to the noninformation resource they want to talk about; and the other to provide access to information describing it.

But knowing the first, how can the second one be found? The cool URI for the Semantic Web technique (Sauermann and Cyganiak 2008) solves this problem in a smart and elegant fashion using the very same HTTP protocol. It makes the URIRef identifying a noninformation resource dereferenceable and uses two alternative strategies to enable a HTTP server to answer a request containing that URIRef. The two strategies are called *303 URI* and *hash URI.* With the 303 URI, a client sends a HTTP request to the URIRef associated with the noninformation resource. In the Accept header of the HTTP request, the client specifies the representation it prefers, text/html for getting back a human-readable description (i.e., a web page) and application/rdf+xml or application/rdf+n3 to obtain an RDF representation of the noninformation resource in RDF/XML or N3, respectively. The request is routed to the responding HTTP server, and the server replies with a HTTP containing:

- A 303 REDIRECT status code
- A HTTP Content-Location header specifying the HTTP URI of an information resource describing the resource

Upon receiving this response, the client issues a second HTTP request to the URI specified in the Content-Location header and finally retrieves the requested representation.

A second strategy is to use hash URIs. This technique allows identification of a secondary resource by reference to a primary resource, a tactic traditionally used by web browsers to display a portion of a Hypertext Markup Language (HTML) document—one identified by the fragment—to the user. For example, when the following URIRef is put in the browser bar

```
http://example.org/myPage.html#myFragment
```

the whole web page http://example.org/myPage.html is requested to the web server. Then the browser locally processes the page and displays the portion corresponding to `#myFragment` (RFC 3986; Berners-Lee et al. 2005).

Hash URIs reuse this idea; the fragment identifier of a URIRef distinguishes the noninformation resource from the corresponding information resource. The URIRef of the latter is simply given by the URIRef of the former without the fragment identifier (so-called base URI). Because the fragment identifier is stripped by the HTTP client before sending the HTTP request, the issued request is to an information resource even if the original URIRef refers to a noninformation resource.

Processing the fragment identifier is exclusively client-sided and so allows the client to retrieve a representation by issuing only one HTTP request instead of two (as in 303 URIs). Unfortunately, many noninformation resources can share the same base URI, so the size of the retrieved document and its accommpaning descriptions may be copious.

In either case, cool URIs should be built to make resources as discoverable as possible enabling humans to "follow their nose" (an expression used in the linked data community) to find more information related to their identified resource. Hash URIs obviously honor this recommendation as it is enough to strip the fragment to retrieve the corresponding representation. Regarding 303 URI, it is common to use consistently mnemonic strings to distinguish the URIRef identifying the noninformation resource from the ones locating its HTML or RDF representations. For instance, an imaginary site, example.org could use the following pattern (Heath and Bizer 2011):

```
http://example.org/resource/anyResource
http://example.org/html/anyResource
http://example.org/rdf/anyResource
```

to identify all its resources. A human being can then unequivocally identifiy which URIRef refers to the noninformation resource, which is assigned to its corresponding representations, and guess the latter from the former. It is important to note however that this is a convention established to make references more human readable but that software clients must not rely on it to distinguish how the three URIRefs are used. The URIRefs are always opaque (Jacobs and Walsh 2004) to the client. The only mechanism a client should be aware of is the 303 REDIRECT technique previously described, which is consistent with the REST practice of using the semantics of HTTP headers to provide additional information during the communication.

A second recommendation is that cool URIs should avoid port numbers, queries, and Internet Protocol (IP) addresses. This ensures more persistent references because it abstracts as much as possible from the implementation details. Cool URIs ideally should simply contain the domain name, the target resource's name, and possibly an indication of the kind of resource identified (a noninformation resource, an HTML page, or an RDF description).

Linking Data Sets

To work, the cool URIs technique assumes that a URIRef must be dereferenceable (or redirectable), must return a representation when dereferenced, and must be reasonably persistent over time. Because only the URI owner can ensure that these requirements are met, data publishers should create URIRefs only from their own domain. But what happens when two publishers reference the same entity? Which URIRefs should they use?

Because the open world of linked data is totally decentralized, there are no "official URIRefs" for any entity. Each publisher should use its own and possibly set equivalence links by other data publishers for some. This process is vital for linked data because it is the primary method to enable integration of different datasets from different sources.

Establishing an equivalence link is in principle, a very simple task. OWL (Web Ontology Language) provides a property, `owl:sameAs`, that allows the data publisher to assert equivalences. To say that two URIRefs usually from different domains refer to the same individual,* it is sufficient but necessary to state that they are in an `owl:sameAs` relationship, that is,

```
@prefix owl: <http://www.w3.org/2002/07/owl#>.
@prefix ex1: <http://example-one.org/resources#>.
@prefix ex2: <http://example-two.org/resources#>.

ex1:aReference owl:sameAs ex2:anotherReference.
```

`ex:ref1` and `ex:ref2` are said to be co-references.

* In OWL DL and OWL Light, the property `owl:sameAs` applies to instances only (classes and properties are related using `owl:equivalentClass` and `owl:equivalentProperty`).

Being `owl:sameAs`, in a symmetric relationship, the inverse statement should hold as well:

```
@prefix owl: <http://www.w3.org/2002/07/owl#>.
@prefix ex1: <http://example-one.org/resources#>.
@prefix ex2: <http://example-two.org/resources#>.

ex1:aReference owl:sameAs ex2:anotherReference.
ex2:anotherReference owl:sameAs ex1:aReference.
```

The formal semantics of `owl:sameAs` is very rigid, and great care should be taken before publishing equivalences. In fact, to conform with the semantics of `owl:sameAs` the coreferences should be *completely* interchangeable. If one resource has certain properties, these should be true for the other one and vice versa. Even if this seems obvious, a recent study has highlighted that `owl:sameAs` has many varying concrete usages, some not totally compliant with its formal semantics. Halpin et al. (2010) present four alternative readings of how this property is used: (1) misplaced references (using an inverse-functional property like an email address for a reference to an entity, such as a person); (2) referential opacity (mostly using a reference without caring if some of the properties were originally defined for it); (3) identity in different contexts (not considering that the same person might have different properties at different ages or places or in different organizations); and (4) similarity (a weaker relationship than equivalence). None of these cases respects the original semantics!

The Simple Knowledge Organization System vocabulary (Miles and Bechhofer 2009) defines alternative properties that might better replace `owl:sameAs` in the aforementioned cases, but choosing suitable alternatives might prove to be a difficult subjective task because someone's close match may be another person's identical match (Halpin et al. 2010). Thus, community consensus becomes fundamental and appears to suggest that the task of evaluating an equivalence link as correct or incorrect requires a global perspective rather than a local decision by a single data publisher. A bird's-eye view of the entire set of coreferences and their equivalence links may provide more useful indications for the strength of a given equivalence than will looking at each equivalence statement individually. As a matter of fact, the linked data cloud has begun to show widely referred nodes attracting a large portion of incoming equivalence links, becoming

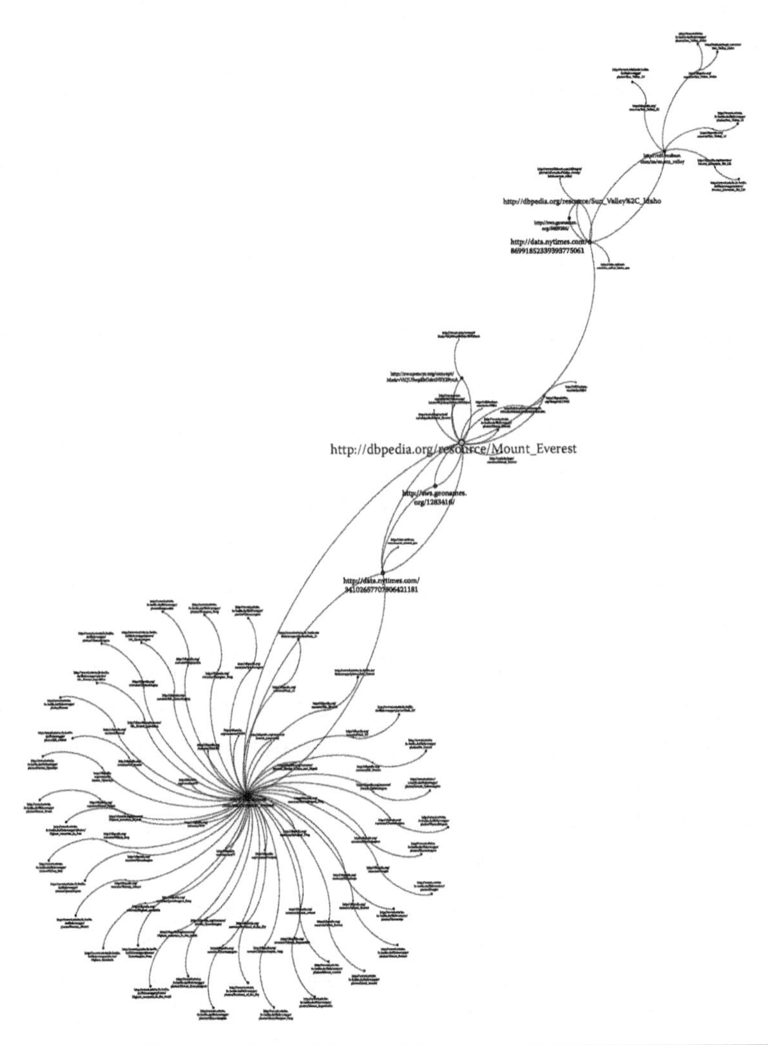

Figure 22.1 An RDF graph made of only equivalence links (data collected from Linked Open Data cloud).

for the linked data community more representative of an entity than others. This illustrates at least three particular features: distribution, because for each entity more than one node might aspire to become an "authority"; dynamism, because it varies over time according to the rise of new pay-level domains* publishing their entity descriptions as linked data; and aggregation, because groups of coreferences coming from different domains tend to cluster. Figure 22.1 represents a

* Pay-level domain is the term used to identify a domain subordinate to a generic top-level domain or to a country-code top-level domain.

characteristic RDF graph made of equivalence links only. The graph has been built by recursively following the `owl:sameAs` connections between nodes from different domains beginning at the resource <http://dbpedia.org/resource/Mount_Everest>. Cluster analysis reveals that there are two different pairs of clusters. The one at bottom left is formed by numerous equivalence links from freebase.com, which establishes coreferences between the resource <http://rdf.freebase.com/ns/en.mount_everest> and resources in dbpedia.org. The second one develops around the central resource <http://dbpedia.org/resource/Mount_Everest> and accounts for the many links established by dbpedia.com with other well-known domains. Finally, the two on the top right refer to a different entity entirely (Sun Valley, Idaho). The graph actually reveals mistakenly added equivalence links proving that providing even a single wrong equivalence link could lead to dangerous unexpected behaviors in software systems.

References

3rd Generation Partnership Project. 2000. Technical Specification Group Services and System Aspects; 3G Security; Security Architecture. 3GPP TS 33.102.

3rd Generation Partnership Project. 2004a. Generic Authentication Architecture (GAA); Access to Network Application Functions Using Hypertext Transfer Protocol over Transport Layer Security (HTTPS). 3GPP TS 33.222.

3rd Generation Partnership Project. 2004b. Technical Specification Group Core Network and Terminals; Bootstrapping Interface (Ub) and Network Application Function Interface (Ua); Protocol Details. 3GPP TS 24.109.

3rd Generation Partnership Project. 2004c. Technical Specification Group Services and System Aspects; 3G Security; Generic Authentication Architecture (GAA); Generic Bootstrapping Architecture. 3GPP TS 33.220.

3rd Generation Partnership Project. 2004d. Technical Specification Group Services and System Aspects; 3G Security; Generic Authentication Architecture (GAA); Support for Subscriber Certificates. 3GPP TS 33.221.

3rd Generation Partnership Project. 2011. Technical Specification Group Services and System Aspects; 3G Security; Generic Authentication Architecture (GAA); System Description (Release 10). 3GPP TR 33.919.

Ahlgren, B., D'Ambrosio, M., Marchisio, M. et al. 2008. Design Considerations for a Network of Information. In *Proceedings of the 2008 ACM CoNEXT Conference*, Madrid, Spain, December 9–12, edited by Azcorra, A., de Veciana, G., Ross, K. W. et al. New York: ACM.

Alves, A., Arkin, A., Askary, S. et al. 2007. Web Services Business Process Execution Language Version 2.0. OASIS Standard April 11, 2007. http://docs.oasis-open.org/wsbpel/2.0/wsbpel-v2.0.pdf.

Anklesaria, F., McCahill, M., Lindner, P., Johnson, D., Torrey, D., and Albert, B. 1993. The Internet Gopher Protocol (a Distributed Document Search and Retrieval Protocol). RFC 1436, March.

Arenas, A., Bertails, A., Prud'hommeaux, E. et al. (eds.). 2012. A Direct Mapping of Relational Data to RDF W3C Recommendation, September 27. World Wide Web Consortium. http://www.w3.org/TR/rdb-direct-mapping/.

Balakrishnan, H., Kaashoek, M., Karger, D. et al. 2003. Looking Up Data in P2P Systems. *Communications of the ACM* 46(2): 43–48.

Becker, G. 2008. Merkle Signature Schemes, Merkle Trees and Their Cryptanalysis. Seminar "Post Quantum Cryptology" at Ruhr University, Bochum, Germany.

Beckett D., and Berners-Lee, T. (eds.). 2011. Turtle—Terse RDF Triple Language. W3C Team Submission, March 28, 2011. World Wide Web Consortium. http://www.w3.org/TeamSubmission/turtle/.

Beckett, D. (ed.). 2004. RDF/XML Syntax Specification (Revised). W3C Recommendation, February 10. World Wide Web Consortium. http://www.w3.org/TR/rdf-syntax-grammar/.

Berners-Lee, T. 1998. Cool URIs Don't Change. http://www.w3.org/Provider/Style/URI.

Berners-Lee, T. 2006. Linked Data. http://www.w3.org/DesignIssues/LinkedData.html.

Berners-Lee, T. 2010. The Future of RDF. Paper presented at W3C Workshop—RDF Next Steps, Stanford, Palo Alto, CA, June 26–27. http://www.w3.org/DesignIssues/RDF-Future.html.

Berners-Lee, T. 1994. Universal Resource Identifiers in WWW: A Unifying Syntax for the Expression of Names and Addresses of Objects on the Network as Used in the World-Wide Web. RFC 1630, June.

Berners-Lee, T., Fielding, R., and L. Masinter. 1998. Uniform Resource Identifiers (URI): Generic Syntax. RFC 2396, August.

Berners-Lee, T., Fielding, R., and L. Masinter. 2005. Uniform Resource Identifier (URI): Generic Syntax. STD 66, RFC 3986, January.

Berners-Lee, T., Masinter, L., and M. McCahill. 1994. Uniform Resource Locators (URL). RFC 1738, December.

Blefari Melazzi, N., Salsano, S., Detti, A. et al. 2012. Publish/Subscribe over Information Centric Networks: A Standardized Approach in CONVERGENCE. In *Proceedings of the Future Network & Mobile Summit* (FutureNetw) 2012. Berlin, Germany, July 4–6.

Boag, S., Chamberlin, D., Fernández, M. F. et al. (eds.). 2010. XQuery 1.0: An XML Query Language (Second Edition). W3C Recommendation, December 14. World Wide Web Consortium. http://www.w3.org/TR/xquery/.

Borenstein, N., and Freed, N. 1993. MIME (Multipurpose Internet Mail Extensions) Part One: Mechanisms for Specifying and Describing the Format of Internet Message Bodies. RFC 1521, September.

Bradner, S., Conroy, L., and Fujiwara, K. 2011. The E.164 to Uniform Resource Identifiers (URI) Dynamic Delegation Discovery System (DDDS) Application (ENUM). RFC 6116, March.

Brickley, D. 2011. Dilbert example—defining hasCubicle. Email message to W3C RDF Working Group, October 13. http://lists.w3.org/Archives/Public/public-rdf-wg/2011Oct/0232.html.

Brickley, D., and Guha, R. V. (eds.). 2004. RDF Vocabulary Description Language 1.0: RDF Schema. W3C Recommendation, February 10. World Wide Web Consortium. http://www.w3.org/TR/rdf-schema/.

Caesar, M., Condie, T., Kannan, J. et al. 2006. ROFL: Routing on Flat Labels. In *Proceedings of the 2006 Conference on Applications, Technologies, Architectures, and Protocols for Computer Communications,* Pisa, Italy, September 11–15, edited by Rizzo, L., Anderson, T., and McKeown, N., 363–374. New York: ACM.

Cafarella, M. J., Madhavan, J., and Halevy, A. 2009. Web-Scale Extraction of Structured Data. *ACM SIGMOD Record Archive* 37(4): 55–61.

Calhoun, P., Loughney, J., Guttman, E., Zorn, G., and Arkko, J. 2003. Diameter Base Protocol. RFC 3588, September.

Campbell, B. (ed.), Rosenberg, J., Schulzrinne, H., Huitema, C., and Gurle, D. 2002. Session Initiation Protocol (SIP) Extension for Instant Messaging. RFC 3428, December.

Carpenter, B. 2000. Internet Transparency. RFC 2775, February.

Carroll, J. J., Bizer, C., Hayes, P. et al. 2005. Named Graphs. *Journal of Web Semantics: Science, Services and Agents on the World Wide Web* 3(4): 247–267.

Christensen, E., Curbera, F., Meredith, G. et al. (eds.) 2001. Web Services Description Language (WSDL) 1.1 W3C Note 15, March. World Wide Web Consortium. http://www.w3.org/TR/wsdl.

Clark, J., and DeRose, S. (eds.) 1999. XML Path Language (XPath) Version 1.0. W3C Recommendation, November 16. World Wide Web Consortium. http://www.w3.org/TR/xpath/.

Clement, L., Hately, A., von Riegen, C. et al. 2004. UDDI Version 3.0.2. UDDI Spec Technical Committee Draft, Dated 20041019. https://www.oasis-open.org/committees/uddi-spec/doc/spec/v3/uddi-v3.0.2-20041019.htm.

Connolly, D. (ed.). 2007. Gleaning Resource Descriptions from Dialects of Languages (GRDDL). W3C Recommendation, September 11. World Wide Web Consortium. http://www.w3.org/TR/grddl/.

Connolly, D., van Harmelen, F., Horrocks, I. et al. (eds.). 2001. DAML+OIL (March 2001) Reference Description. W3C Note, December 18. World Wide Web Consortium. http://www.w3.org/TR/daml+oil-reference.

Cooper, I., Melve, I., and Tomlinson, G. 2001. Internet Web Replication and Caching Taxonomy. RFC 3040, January.

Cotton, M., and Vegoda, L. 2010. Special Use IPv4 Addresses. BCP 153, RFC 5735, January.

Court, G. 2010. A JSON Media Type for Describing the Structure and Meaning of JSON. IETF Internet-Draft draft-zyp-json-schema-03. http://tools.ietf.org/html/draft-zyp-json-schema-03.

Crawford, M., and Huitema, C. 2000. DNS Extensions to Support IPv6 Address Aggregation and Renumbering. RFC 2874, July.

Crockford, D. 2006. The Application/JSON Media Type for JavaScript Object Notation (JSON). RFC 4627, July.

Cyganiak, R., Harth, A., and Hogan, Quads, A.-N. 2008. Extending N-Triples with Context. http://sw.deri.org/2008/07/n-quads/.

Daniel, R., and Mealling, M. 1997. Resolution of Uniform Resource Identifiers Using the Domain Name System. RFC 2168, June.

Das, S., Sundara, S., and Cyganiak, R. (eds.). 2012. R2RML: RDB to RDF Mapping Language W3C Recommendation, September 27. World Wide Web Consortium. http://www.w3.org/TR/r2rml/.

De Rose, S., Daniel, R., Jr., Grosso, P. et al. (eds.). 2011. XML Pointer Language (XPointer). World Wide Web Consortium. http://www.w3.org/TR/xptr/.

Deering, S. 1989. Host Extensions for IP Multicasting. STD 5, RFC 1112, August.

Deering, S. 2001. Watching the Waist of the Protocol Hourglass. Keynote at the IETF 51st Meeting, London, August 5–10. http://www.iab.org/iab/DOCUMENTS/hourglass-london-ietf.pdf.

DeRose, S., Maler, E., and Orchard, D. (eds.). 2001. XML Linking Language (XLink) Version 1.0. W3C Recommendation, June 27. World Wide Web Consortium. http://www.w3.org/TR/xlink/

Dierks, T., and Allen, C. 1999. The TLS Protocol Version 1.0. RFC 2246, January.

Dierks, T., and Rescorla, E. 2006. The Transport Layer Security (TLS) Protocol Version 1.1. RFC 4346, April.

Ding, L., Finin, T., Peng, Y. et al. 2005. Tracking RDF Graph Provenance Using RDF Molecules. Technical Report TR-05-06, University of Maryland, Baltimore.

Duerst, M., and Suignard, M. 2005. Internationalized Resource Identifiers (IRIs). RFC 3987, January.

Dumbill, E. 2003. XML Watch: Tracking Provenance of RDF Data. http://www.ibm.com/developerworks/xml/library/x-rdfprov/index.html.

Dusseault, L., and Snell, J. 2010. PATCH Method for HTTP. RFC 5789, March.

Eastlake, D., III, and Panitz, A. 1999. Reserved Top Level DNS Names. BCP 32, RFC 2606, June.

Emtage, A., and Deutsch, P. 1992. Archie—An Electronic Directory Service for the Internet. *Proceedings of the Winter USENIX Conference*, Usenix Association, Berkeley, San Francisco, CA, 93–110.

Eronen, P., and Tschofenig, H. (eds.). 2005. Pre-Shared Key Ciphersuites for Transport Layer Security (TLS). RFC 4279, December.

European Telecommunications Standards Institute (ETSI). 2006a. PSTN/ISDN Emulation Sub-system (PES); Functional Architecture. ETSI ES 282 002. Sophia Antipolis, France: Author.

European Telecommunications Standards Institute (ETSI). 2008a. Telecommunications and Internet converged Services and Protocols for Advanced Networking (TISPAN): IP Multimedia Subsystem (IMS): Functional Architecture. ETSI ES 282 007. Sophia Antipolis, France: Author.

European Telecommunications Standards Institute (ETSI). 2000. Digital Cellular Telecommunications System (Phase 2+); Universal Mobile Telecommunications System (UMTS); Numbering, Addressing and Identification. ETSI TS 123 003. Sophia Antipolis, France: Author.

European Telecommunications Standards Institute (ETSI). 2001. Human Factors (HF); User Identification Solutions in Converging Networks. ETSI EG 201 940. Sophia Antipolis, France: Author.
European Telecommunications Standards Institute (ETSI). 2002. Universal Communications Identifier (UCI); System Framework. ETSI EG 202 067. Sophia Antipolis, France: Author.
European Telecommunications Standards Institute (ETSI). 2003. Universal Communications Identifier (UCI); Results of a Detailed Study into the Technical Areas for Identification Harmonization; Recommendations on the UCI for NGN. ETSI EG 203 072. Sophia Antipolis, France: Author.
European Telecommunications Standards Institute (ETSI). 2005. Telecommunications and Internet Converged Services and Protocols for Advanced Networking (TISPAN); NGN Functional Architecture. ETSI ES 282 001. Sophia Antipolis, France: Author.
European Telecommunications Standards Institute (ETSI). 2006. Telecommunications and Internet Converged Services and Protocols for Advanced Networking (TISPAN); Identifiers (IDs) for NGN. ETSI TS 184 002. Sophia Antipolis, France: Author.
European Telecommunications Standards Institute (ETSI). 2007. Terrestrial Trunked Radio (TETRA); Voice plus Data (V+D); Part 2: Air Interface (AI). ETSI EN 300 392-2. Sophia Antipolis, France: Author.
European Telecommunications Standards Institute (ETSI). 2008a. Telecommunications and Internet Converged Services and Protocols for Advanced Networking (TISPAN); NGN Security; Report on Issues Related to Security in Identity Management and Their Resolution in the NGN. ETSI TR 187 010. Sophia Antipolis, France: Author.
European Telecommunications Standards Institute (ETSI). 2008b. Telecommunications and Internet Converged Services and Protocols for Advanced Networking (TISPAN); NGN Integrated IPTV Subsystem Architecture. ETSI TS 182 028. Sophia Antipolis, France: Author.
European Telecommunications Standards Institute (ETSI). 2009. Human Factors (HF); Personalization and User Profile Management; User Profile Preferences and Information. ETSI ES 202 746. Sophia Antipolis, France: Author.
European Telecommunications Standards Institute (ETSI). 2010. NGN Functional Architecture; Network Attachment Sub-System (NASS). ETSI ES 282 004. Sophia Antipolis, France: Author.
European Telecommunications Standards Institute (ETSI). 2011a. Resources and Admission Control Sub-System (RACS); Functional Architecture. ETSI ES 282 003. Sophia Antipolis, France: Author.
European Telecommunications Standards Institute (ETSI). 2011b. Telecommunications and Internet converged Services and Protocols for Advanced Networking (TISPAN); IPTV Architecture; IPTV Functions Supported by the IMS Subsystem. ETSI TS 182 027. Sophia Antipolis, France: Author.

European Telecommunications Standards Institute (ETSI). 2011c. Telecommunications and Internet Converged Services and Protocols for Advanced Networks (TISPAN); ENUM & DNS Principles for an Interoperator IP Backbone Network. ETSI TS 184 010. Sophia Antipolis, France: Author.

Farrell, J., and Lausen, H. (eds.). 2007. Semantic Annotations for WSDL and XML Schema. W3C Recommendation, August 28. World Wide Web Consortium. http://www.w3.org/TR/sawsdl/.

Feigenbaum, E. 2004. Keynote at the Third International Semantic Web Conference, Hiroshima, Japan, November 7–11.

Fielding, R. 1995. Relative Uniform Resource Locators. RFC 1808, June.

Fielding, R. T. 2000. Architectural Styles and the Design of Network-Based Software Architectures. Ph.D. diss., University of California, Irvine. http://www.ics.uci.edu/~fielding/pubs/dissertation/top.htm.

Fielding, R., Gettys, J., Mogul, J., Frystyk, H., and Berners-Lee, T. 1997. Hypertext Transfer Protocol—HTTP/1.1. RFC 2068, January.

Fielding, R., Gettys, J., Mogul, J. et al. 1999. Hypertext Transfer Protocol—HTTP/1.1. RFC 2616, June.

Fielding, R. T. 2005. httpRange-14 Resolved. 2005. Email to W3C TAG. June 18. http://lists.w3.org/Archives/Public/www-tag/2005Jun/0039.html.

Franks, J., Hallam-Baker, P., Hostetler, J. et al. 1997. An Extension to HTTP: Digest Access Authentication. RFC 2069, January.

Franks, J., Hallam-Baker, P., Hostetler, J. et al. 1999. HTTP Authentication: Basic and Digest Access Authentication. RFC 2617, June.

Fuller, V., and Li, T. 2006. Classless Inter-domain Routing (CIDR): The Internet Address Assignment and Aggregation Plan. BCP 122, RFC 4632, August.

Gantz, J., and Reinsel, D. 2011. The 2011 Digital Universe Study: Extracting Value from Chaos. IDS Website, June. http://www.emc.com/collateral/demos/microsites/emc-digital-universe-2011/index.htm.

Graham, S., Hull, D. M., and Murray, B. 2006. Web Services Base Notification 1.3 (WS-BaseNotification). OASIS Standard, 1 October. http://docs.oasis-open.org/wsn/wsn-ws_base_notification-1.3-spec-os.pdf.

Grant, J., and Beckett, D. (eds.). 2004. RDF Test Cases. W3C Recommendation, February 10. World Wide Web Consortium. http://www.w3.org/TR/rdf-testcases/.

Gregorio, J., and de hOra, B. (eds.). 2007. The Atom Publishing Protocol. RFC 5023, October.

Gritter, M., and Cheriton, D. R. 2001. An Architecture for Content Routing Support in the Internet. In *Proceedings of the Third Conference on USENIX Symposium on Internet Technologies and Systems,* co-sponsored by IEEE Computer Society and Information Processing Society of Japan (IPSTJ), edited by Prof. Ikeda and Dr. Douglis, progam co-chairs, Vol. 3. Berkeley, CA: USENIX Association, 4–4.

Gudgin, M., Hadley, M., Mendelsohn, N. et al. (eds.). 2007. SOAP Version 1.2 Part 1: Messaging Framework (2nd ed.). W3C Recommendation, April 27. World Wide Web Consortium. http://www.w3.org/TR/soap12-part1/.

Gudgin, M., Hadley, M., and Rogers, T. (eds.). 2006. Web Services Addressing 1.0—Core. W3C Recommendation, May 9. World Wide Web Consortium. http://www.w3.org/TR/ws-addr-core/.

Halpin, H., Hayes, P. P., McCusker, J. et al. 2010. When owl:sameAs Isn't the Same: An Analysis of Identity in Linked Data. In *Proceedings of the Ninth International Semantic Web Conference, Vol. 1,* Shanghai, China, November 7–11, edited by Patel-Schneider, P. F., Pan, Y., Hitzler, P. et al., 305–320. Berlin: Springer-Verlag.

Handley, M., Jacobson, V., and Perkins, C. 2006. SDP: Session Description Protocol. RFC 4566, July.

Handley, M., Schulzrinne, H., Schooler, E., and Rosenberg, J. 1999. SIP: Session Initiation Protocol. RFC 2543, March.

Harkins, D., and Carrel, D. 1998. The Internet Key Exchange (IKE). RFC 2409, November.

Harren, M., Joseph, M. H., and Huebsch, R. 2002. Complex Queries in DHT-Based Peer-to-Peer Networks. In *Proceeding of The First International Workshop on Peer-to-Peer Systems,* Cambridge, MA, March 7–8, edited by Goos, G., Hartmanis, J., and van Leeuwen, J., 242–250. London: Springer-Verlag.

Hausenblas, M. 2010. Five Stars Model on Open Government Data. http://lab.linkeddata.deri.ie/2010/star-scheme-by-example.

Hayes, P. 2011. Graph Names as Third Argument. E-mail to W3C RDF Workgroup, November 2. http://lists.w3.org/Archives/Public/public-rdf-wg/2011Nov/0019.html.

Hayes, P., and Halpin, H. 2008. In Defense of Ambiguity. *International Journal of Semantic Web and Information Systems* 4(2): 1–18.

Heath, T., and Bizer, C. 2011. *Linked Data: Evolving the Web into a Global Data Space. Synthesis Lectures on the Semantic Web: Theory & Technology.* San Francisco, CA: Morgan Claypool. http://www.morganclaypool.com/page/aboutMCp.jsp.

Hickson, I. 2010. The WebSocket Protocol V76, IETF Internet-Draft. Last updated May 6. http://tools.ietf.org/html/draft-hixie-thewebsocketprotocol-76.

Hinden, R. 1994. Simple Internet Protocol Plus White Paper. RFC 1710, October.

Hofmann, M., and Beaumont, L. R. 2005. *Content Networking: Architecture, Protocols, and Practice.* San Francisco, CA: Morgan Kaufmann.

International Telecommunication Union (ITU) Telecommunication Standardization Bureau (TSB). 1992. Interface between Data Terminal Equipment and Data Circuit-Terminating Equipment for Synchronous Operation On Public Data. ITU-T Recommendation X.21. Geneva, Switzerland: Author.

International Telecommunication Union (ITU) Telecommunication Standardization Bureau (TSB). 1996. Interface between Data Terminal Equipment (DTE) and Data Circuit Terminating Equipment (DCE) for Terminals Operating in the Packet Mode and Connected to Public Data Networks by Dedicated Circuit. ITU-T Recommendation X.25. Geneva, Switzerland: ITU.

International Telecommunication Union (ITU) Telecommunication Standardization Bureau (TSB). 1999. Assignment Procedures for International Signalling Point Codes. ITU-T Recommendation Q.708. Geneva, Switzerland: ITU.

International Telecommunication Union (ITU) Telecommunication Standardization Bureau (TSB). 2000a. B-ISDN Addressing. ITU-T Recommendation E.191. Geneva, Switzerland: ITU.

International Telecommunication Union (ITU) Telecommunication Standardization Bureau (TSB). 2000b. International Numbering Plan for Public Data Networks. ITU-T Recommendation X.121. Geneva, Switzerland: Author.

International Telecommunication Union (ITU) Telecommunication Standardization Bureau (TSB). 2005. Infrastructure of Audiovisual Services—Communication Procedures Gateway Control Protocol. ITU-T Recommendation H.248.1. Geneva, Switzerland: Author.

International Telecommunication Union (ITU) Telecommunication Standardization Bureau (TSB). 2006. The International Telecommunication Charge Card. ITU-T Recommendation E.118. Geneva, Switzerland: ITU.

International Telecommunication Union (ITU) Telecommunication Standardization Bureau (TSB). 2008. The International Identification Plan for Public Networks and Subscriptions. ITU-T Recommendation E.212. Geneva, Switzerland: ITU.

International Telecommunication Union (ITU) Telecommunication Standardization Bureau (TSB). 2009. Infrastructure of Audiovisual Services—Systems and Terminal Equipment for Audiovisual Services Packet-Based Multimedia Communications Systems. ITU-T Recommendation H.323. Geneva, Switzerland: Author.

International Telecommunication Union (ITU) Telecommunication Standardization Bureau (TSB). 2010. The International Public Telecommunication Numbering Plan. ITU-T Recommendation E.164. Geneva, Switzerland: ITU.

Jacobs, I., and Walsh, N. (eds.). 2004. Architecture of the World Wide Web, Volume One. W3C Recommendation, December 15. World Wide Web Consortium. http://www.w3.org/TR/webarch/.

Jacobson, V., Smetters, D. K., Briggs, N. et al. 2009a. VoCCN: Voice over Content-Centric Networks. In *Proceedings of the 2009 Workshop on Re-Architecting the Internet,* Rome, Italy, December 1, edited by Eggert, L., and Wolf, T., 1–6. New York: ACM.

Jacobson, V., Smetters, D. K., Thornton, J. D. et al. 2009b. Networking Named Content. *Communications of the ACM* 55(1): 117–124.

Jennings, C., Peterson, J., and Watson, M. 2002. Private Extensions to the Session Initiation Protocol (SIP) for Asserted Identity within Trusted Networks. RFC 3325, November.

Kahle, B., and Medlar, A. 1991. An Information System for Corporate Users: WAIS. *ConneXions—The Interoperability Report* 5(11): 2–9.

Klyne, G., Carroll, J. J., and McBride, B. (eds.). 2012. RDF 1.1 Concepts and Abstract Syntax. W3C Working Draft, June 5. World Wide Web Consortium. http://www.w3.org/TR/rdf11-concepts/#xsd-datatypes.

Koponen, T., Chawla, M., Chun, B. G. et al. 2007. A Data-Oriented (and beyond) Network Architecture. *ACM SIGCOMM—Computer Communication Review* 37(4): 181–192.

Maymounkov, P., and Mazieres, D. 2002. Kademlia: A peer-to-peer information system based on the xor metric. In *Proceeding of The First International Workshop on Peer-to-Peer Systems,* Cambridge, MA, March 7–8, edited by Goos, G., Hartmanis, J., and van Leeuwen, J., 53–65. London: Springer-Verlag.

McGuinness, D. L., and van Harmelen, F. (eds.). 2004. OWL Web Ontology Language. Overview. W3C Recommendation, February 10. World Wide Web Consortium. http://www.w3.org/TR/owl-features/.

Miles, A., and Bechhofer, S. (eds.). 2009. SKOS Simple Knowledge Organization System Reference. W3C Recommendation, August 18. World Wide Web Consortium. http://www.w3.org/TR/2009/REC-skos-reference-20090818/.

Moats, R. 1997. URN Syntax. RFC 2141, May.

Mockapetris, P. 1987a. Domain Names—Concepts and Facilities. STD 13, RFC 1034, November.

Mockapetris, P. 1987b. Domain Names—Implementation and Specification. STD 13, RFC 1035, November.

Motik, B. 2005. On the Properties of Metamodeling in OWL. In *Proceeding of the Fourth International Semantic Web Conference,* Galway, Ireland, November 6–10, edited by Gil, Y., and Motta, E., 548–562. Berlin: Springer-Verlag.

Niemi, A., Arkko, J., and Torvinen, V. 2002. Hypertext Transfer Protocol (HTTP) Digest Authentication Using Authentication and Key Agreement (AKA), RFC 3310, September.

Nocentini, C., Crescenzi P., and Lanzi, L. 2009. Performance Evaluation of a Chord-Based JXTA Implementation. In *Advances in P2P Systems, 2009. AP2PS '09. First International Conference on,* Sliema, Malta, October 11–16, edited by Liotta, A., Antonopoulos, N., Exarchakos, G., and Hara, T., 7–12. Washington, DC: IEEE Computer Society.

Nottingham, M., and Sayre, R. (eds.). 2005. The Atom Syndication Format. RFC 4287, December.

O'Hearn, "Zooko" W. 2006. Names: Decentralized, Secure, Human-Meaningful: Choose Two. http://zooko.com/distnames.html.

Parsons, G., and Rafferty, J. 1998. Tag Image File Format (TIFF)—F Profile for Facsimile. RFC 2306, March.

Paterson, I., Smith, D., Saint-Andre, P. et al. 2010. XEP-0124: Bidirectional-Streams Over Synchronous HTTP (BOSH). Last updated July 2, 2010. http://xmpp.org/extensions/xep-0124.html.

Pavel, S. 2010. Programming SIP Services—The SIP APIs. *Acta Electrotechnica et Informatica* 10(4): 39–45.

Pizzo, M. 2012. Basic OData Question. E-Mail Message to OASIS OData Technical Committee, October 7. https://lists.oasis-open.org/archives/odata/201210/msg00057.html.

Postel, J. 1981. Internet Protocol. STD 5, RFC 791, September.

Postel, J., and Reynolds, J. 1985. File Transfer Protocol. STD 9, RFC 959, October.

Rekhter, Y., Moskowitz, B., Karrenberg, D., de Groot, G., and Lear, E. 1996. Address Allocation for Private Internets. BCP 5, RFC 1918, February.

Rescorla, E. 2000. HTTP over TLS. RFC 2818, May.

Roach, A. 2002. Session Initiation Protocol (SIP)-Specific Event Notification. RFC 3265, June.

Roach, A., Campbell, B., and Rosenberg, J. 2006. A Session Initiation Protocol (SIP) Event Notification Extension for Resource Lists. RFC 4662, August.

Rosenberg, J. 2007a. The Extensible Markup Language (XML) Configuration Access Protocol (XCAP). RFC 4825, May.

Rosenberg, J. 2007b. Extensible Markup Language (XML) Formats for Representing Resource Lists. RFC 4826, May.

Rosenberg, J., and Schulzrinne, H. 2002a. An Offer/Answer Model with Session Description Protocol (SDP). RFC 3264, June.

Rosenberg, J., and Schulzrinne, J. 2002b. Reliability of Provisional Responses in Session Initiation Protocol (SIP). RFC 3262, June.

Rosenberg, J., and Schulzrinne, H. 2002c. Session Initiation Protocol (SIP): Locating SIP Servers. RFC 3263, June.

Rosenberg, J., Schulzrinne, H., Camarillo, G. et al. 2002. SIP: Session Initiation Protocol. RFC 3261, June.

Rosenberg, J., and Urpalainen, J. 2010. An Extensible Markup Language (XML) Document Format for Indicating a Change in XML Configuration Access Protocol (XCAP) Resources. RFC 5874, May.

Rowstron, A., and Druschel, P. P. 2001. Pastry: Scalable, Distributed Object Location and Routing for Large Scale Peer-to-Peer Systems. In *Proceedings of the IFIP/ACM International Conference on Distributed Systems Platforms,* Heidelberg, Germany, November 12–16, edited by Guerraoui, R., 329–350. London: Springer-Verlag.

RSA Laboratories. 2000. PKCS #10; Certification Request Syntax Standard. http://www.rsasecurity.com/rsalabs/node.asp?id = 2132.

Russell, A., Wilkins, G., Davis, D. et al. 2007. Bayeux Protocol—Bayeux 1.0.0. http://svn.cometd.com/trunk/bayeux/bayeux.html.

Saint-Andre, P. (ed.). 2004. Extensible Messaging and Presence Protocol (XMPP): Core. RFC 3920, October.

Saltzer, J. H., Reed, D. P., and Clark, D. D. 1984. End-to-End Arguments in System Design. *ACM Transactions on Computer Systems (TOCS)* 2(4): 277–288.

SAP News Desk. 2005. Microsoft, IBM, SAP to Discontinue UDDI Web Services Registry Effort. *SOA World Magazine.* http://soa.sys-con.com/node/164624.

Sauermann, L., and Cyganiak, R. (eds.). 2008. Cool URIs for the Semantic Web. W3C Interest Group Note, December 3. World Wide Web Consortium. http://www.w3.org/TR/cooluris/.

Savetz, K., Randall, N., and Lepage, Y. 1996. *Mbone: Multicasting Tomorrow's Internet*. New York: John Wiley & Sons.

Schulzrinne, H., Tschofenig, H., Morris, J., Cuellar, J., Polk, J., and Rosenberg, J. 2007. Common Policy: A Document Format for Expressing Privacy Preferences. RFC 4745, February.

Segec, P., and Kováčiková, T. 2011. *A Survey of Open Source Products for Building a SIP Communication Platform. Advances in Multimedia*. New York, NY: Hindawi Publishing Corporation.

Sollins, K., and Masinter, L. 1994. Functional Requirements for Uniform Resource Names. RFC 1737, December.

Sparks, R. 2003. The Session Initiation Protocol (SIP) Refer Method. RFC 3515, April.

Stiegler, M. 2005. An Introduction to Petname Systems. http://www.skyhunter.com/marcs/petnames/IntroPetNames.html.

Stoica, I., Morris, R., Karger, D. et al. 2001. Chord: A Scalable Peer-to-Peer Lookup Service for Internet Applications. In *Proceedings of the 2001 Conference on Applications, Technologies, Architectures, and Protocols for Computer Communications*, edited by Cruz, R., and Varghese, G., 149–160. New York: ACM.

Tanenbaum, A. S. 2007. ***Computer Networks*** (4th ed.). Upper Saddle River, NJ: Prentice Hall.

Tarkoma, S., Ain, M., and Visala, K. 2009. The Publish/Subscribe Internet Routing Paradigm (PSIRP): Designing the Future Internet Architecture. Helsinki Institute for Information Technology (HIIT), 02015 TKK, Finland.

Thompson, H., S., Beech, D., Maloney, M. et al. (eds.) 2004. XML Schema Part 1: Structures (2nd Ed.). W3C Recommendation, October 28. World Wide Web Consortium. http://www.w3.org/TR/xmlschema-1/.

Unicode Consortium. 2012. The Unicode Standard, Version 6.2.0. http://www.unicode.org/versions/Unicode6.2.0/.

Urpalainen, J., and Willis, D. (eds.). 2010. An Extensible Markup Language (XML) Configuration Access Protocol (XCAP) Diff Event Package. RFC 5875, May.

Vixie, P. 1996. A Mechanism for Prompt Notification of Zone Changes (DNS NOTIFY). IETF RFC 1996, August. http://www.ietf.org/rfc/rfc1996.txt.

Vixie, P. 1999. Extension Mechanisms for DNS (EDNS0). RFC 2671, August.

Wellington, B. 2000. Secure Domain Name System (DNS) Dynamic Update. RFC 3007, November.

Wessels, D., and Claffy, K. 1997a. Application of Internet Cache Protocol (ICP), Version 2. RFC 2187, September.

Wessels, D., and Claffy, K. 1997b. Internet Cache Protocol (ICP), Version 2. RFC 2186, September.

Yergeau, F. 2003. UTF-8, a Transformation Format of ISO 10646. STD 63, RFC 3629, November.

Zeinalipour-Yazti, D., and Folias, T. 2002. *A Quantitative Analysis of the Gnutella Network Traffic*. Riverside: University of California.

Zhang, L., Estrin, D., Burke, J. et al. 2010. Named Data Networking (NDN) Project. http://www.named-data.net/techreport/TR001ndn-proj.pdf.

Index

Z